NORTHEASTERN ILLINOIS UNIVERSITY
3 1224 00397 3467

D0705395

THE IMPLEMENTATION OF THE ICON PROGRAMMING LANGUAGE

PRINCETON SERIES IN COMPUTER SCIENCE
*David R. Hanson and Robert E. Tarjan, Editors*

# The Implementation of the Icon Programming Language

RALPH E. GRISWOLD AND MADGE T. GRISWOLD

PRINCETON UNIVERSITY PRESS

Published by Princeton University Press, 41 William Street,
Princeton, New Jersey 08540

In the United Kingdom: Princeton University Press, Guildford, Surrey





Library of Congress Cataloging in Publication Data will be
found on the last printed page of this book

ISBN 0-691-08431-9

This book has been composed in Times Roman

Clothbound editions of Princeton University Press books are printed on
acid-free paper, and binding materials are chosen for
strength and durability. Paperbacks, although satisfactory
for personal collections, are not usually suitable
for library rebinding





Printed in the United States of America by Princeton University Press
Princeton, New Jersey

## CONTENTS

# PREFACE

Icon is a high-level, general-purpose programming language that offers a broad range of string- and list-processing facilities. It also has a novel expression evaluation mechanism and allows an unusual degree of run-time flexibility. Because of these features, implementing Icon presents problems considerably different from those involved with implementing more traditional languages like Pascal or C.

This book is a study of an implementation of Icon. It differs from the usual books on compilers in emphasizing the implementation of run-time facilities and handling of sophisticated language features.

Icon has been implemented for a wide range of computers, from those running UNIX and VAX/VMS to personal computers running MS-DOS. This is a description of a real implementation in wide use; it is not a toy. The code that appears in the book is real code, as it appears in the source.

Readers of this book should have a general familiarity with programming languages and a general idea of what is involved in implementing a complex software system. Icon is written in C; a reader who has experience with that language will find the extensive examples written in C helpful. A reader who has no C experience, however, can skip the C examples and still grasp the general discussion.

Several groups of readers will find this book especially useful. One is the community of individual designers and experimental implementors who want to implement a language like Icon or to extend an existing implementation of Icon. Another group includes the growing number of sophisticated personal computer users who wish to explore languages like Icon. A third group is the academic community concerned with programming-language implementation techniques. In this context, the book is useful for personal research as well as for courses in programming-language implementation issues and techniques, advanced compiler design, and in-depth studies of high-level programming languages and their implementations.

This book does not attempt to cover all aspects of the implementation of Icon. Instead, it concentrates on central issues and on the more interesting and novel portions of the implementation. Persons who are interested in the details of the implementation, and especially those who want to modify the implementation, will find the source code to be a valuable adjunct to this book. This source code is in the public domain and is readily available. It is important to have the correct version, 6.2, for use with this book, since there are other versions that are considerably different.

Requests for Icon and its source code or for other information about Icon may be directed to:

Icon Project
Department of Computer Science
The University of Arizona
Tucson, Arizona 85721

# ACKNOWLEDGMENTS

The implementation of Icon described in this book owes much to previous work, and in particular to implementations of earlier versions of Icon. Major contributions were made by Cary Coutant, Dave Hanson, Tim Korb, Bill Mitchell, and Steve Wampler. Walt Hansen, Rob McConeghy, and Janalee O'Bagy also made significant contributions to this work.

The present system has benefited greatly from persons who have installed Icon on a variety of machines and operating systems. Rick Fonorow, Bob Goldberg, Chris Janton, Mark Langley, Rob McConeghy, Bill Mitchell, Janalee O'Bagy, John Polstra, Gregg Townsend, and Cheyenne Wills have made substantial contributions in this area.

The support of the National Science Foundation under Grants MCS75-01397, MCS79-03890, MCS81-01916, DCR-8320138, DCR-8401831, and DCR-8502015 was instrumental in the original conception of Icon and has been invaluable in its subsequent development.

A number of persons contributed to this book. Dave Gudeman, Dave Hanson, Bill Mitchell, Janalee O'Bagy, Gregg Townsend, and Alan Wendt contributed to the exercises that appear at the ends of chapters and the projects given in Appendix E. Kathy Cummings, Bill Griswold, Bill Mitchell, Katie Morse, Mike Tharp, and Gregg Townsend gave the manuscript careful readings and made numerous suggestions. Janalee O'Bagy not only read the manuscript but also supplied concepts for presenting and writing the material on expression evaluation.

Finally, Dave Hanson served as an enthusiastic series editor for this book. His perceptive reading of the manuscript and his supportive and constructive suggestions made a significant contribution to the final result.

*Ralph E. Griswold and Madge T. Griswold*

THE IMPLEMENTATION OF THE ICON PROGRAMMING LANGUAGE

CHAPTER 1

# Introduction

PERSPECTIVE: The implementation of complex software systems is a fascinating subject—and an important one. Its theoretical and practical aspects occupy the attention and energy of many persons and it consumes vast amounts of computational resources. In general terms, it is a broad subject ranging from operating systems to programming languages to data-base systems to real-time control systems, and so on.

Past work in these areas has resulted in an increasingly better understanding of implementation techniques, more sophisticated and efficient systems, and tools for automating various aspects of software production. Despite these advances, the implementation of complex software systems remains challenging and exciting. The problems are difficult, and every advance in the state of the art brings new and more difficult problems within reach.

This book addresses a very small portion of the problem of implementing complex software systems—the implementation of a very high-level programming language that is oriented toward the manipulation of structures and strings of characters.

In a narrow sense, this book describes in some detail an implementation of a specific programming language, Icon. In a broader sense, it deals with a language-design philosophy, an approach to implementation, and techniques that apply to the implementation of many programming languages as well as related types of software systems.

The focus of this book is the implementation of programming-language features that are at a high conceptual level—features that are easy for human beings to use as opposed to features that fit comfortably on conventional computer architectures. The orientation of the implementation is generality and flexibility, rather than maximum efficiency of execution. The problem domain is strings and structures rather than numbers. It is these aspects that set the implementation of Icon apart from more conventional programming-language implementations.

## 1.1 IMPLEMENTING PROGRAMMING LANGUAGES

In conventional programming languages, most of the operations that are performed when a program is executed can be determined, statically, by examining the text of the program. In addition, the operations of most programming languages have a fairly close correspondence to the architectural characteristics of the computers on which they are implemented. When these conditions are met, source-code constructs can be mapped directly into machine instructions for the computer on which they are to be executed. The term *compilation* is used for this translation process, and most persons think of the implementation of a programming language in terms of a compiler.

Writing a compiler is a complex and difficult task that requires specialized training, and the subject of compilation has been studied extensively (Waite and Goos 1984; Aho, Sethi, and Ullman 1985). Most of the issues of data representation and code generation are comparatively well understood, and there are now many tools for automating portions of the compiler-writing task (Lesk 1975; Johnson 1975).

In addition to the compiler proper, an implementation of a programming language usually includes a run-time component that contains subroutines for performing computations that are too complex to compile in-line, such as input, output, and mathematical functions.

Some programming languages have features whose meanings cannot be determined statically from the text of a source-language program, but which may change during program execution. Such features include changes in the meaning of functions during execution, the creation of new data types at run time, and self-modifying programs. Some programming languages also have features, such as pattern matching, that do not have correspondences in the architecture of conventional computers. In such cases, a compiler cannot translate the source program directly into executable code. Very high-level operations, such as pattern matching, and features like automatic storage management significantly increase the importance and complexity of the run-time system. For languages with these characteristics—languages such as APL, LISP, SNOBOL4, SETL, Prolog, and Icon—much of the substance of the implementation is in the run-time system rather than in translation done by a compiler. While compiler writing is relatively well understood, run-time systems for most programming languages with dynamic features and very high-level operations are not.

Programming languages with dynamic aspects and novel features are likely to become more important rather than less important. Different problems benefit from different linguistic mechanisms. New applications place different values on speed of execution, memory requirements, quick solutions, programmer time and talent, and so forth. For these reasons, programming languages continue to proliferate. New programming languages, by their nature, introduce new features.

All of this creates difficulties for the implementor. Less of the effort involved in implementations for new languages lies in the comparatively familiar domain of compilation and more lies in new and unexplored areas, such as pattern matching and novel expression-evaluation mechanisms.

The programming languages that are the most challenging to implement are also those that differ most from each other. Nevertheless, there are underlying principles and techniques that are generally applicable, and existing implementations contain many ideas that can be used or extended in new implementations.

## 1.2 THE BACKGROUND FOR ICON

Before describing the Icon programming language and its implementation, some historical context is needed, since both the language and its implementation are strongly influenced by earlier work.

Icon has its roots in a series of programming languages that bear the name SNOBOL. The first SNOBOL language was conceived and implemented in the early 1960s at Bell Telephone Laboratories in response to the need for a programming tool for manipulating strings of characters at a high conceptual level (Farber, Griswold, and Polonsky 1964). It emphasized ease of programming at the expense of efficiency of execution; the programmer was considered to be a more valuable resource than the computer.

This rather primitive language proved to be popular, and it was followed by successively more sophisticated languages: SNOBOL2, SNOBOL3 (Farber, Griswold, and Polonsky 1966), and finally SNOBOL4 (Griswold, Poage, and Polonsky 1971). Throughout the development of these languages, the design emphasis was on ease of programming rather than ease of implementation (Griswold 1981). Potentially valuable features were not discarded because they might be inefficient or difficult to implement. The aggressive pursuit of this philosophy led to unusual language features and to challenging implementation problems.

SNOBOL4 still is in wide use. Considering its early origins, some of its facilities are remarkably advanced. It features a pattern-matching facility with backtracking control structures that effectively constitutes a sublanguage. SNOBOL4 also has a variety of data structures, including tables with associative lookup. Functions and operators can be defined and redefined during program execution. Identifiers can be created at run-time, and a program can even modify itself by means of run-time compilation.

Needless to say, SNOBOL4 is a difficult language to implement, and most of the conventional compilation techniques have little applicability to it. Its initial implementation was, nonetheless, sufficiently successful to make SNOBOL4 widely available on machines ranging from large mainframes to personal computers (Griswold 1972). Subsequent implementations introduced a variety of clever techniques and fast, compact implementations (Santos 1971; Gimpel

1972a; Dewar and McCann 1977). The lesson here is that the design of programming languages should not be overly inhibited by perceived implementation problems, since new implementation techniques often can be devised to solve such problems effectively and efficiently.

It is worth noting that the original implementation of SNOBOL4 was carried out concomitantly with language design. The implementation was sufficiently flexible to serve as a research tool in which experimental language features could be incorporated easily and tested before they were given a permanent place in the language.

Work on the SNOBOL languages continued at the University of Arizona in the early 1970s. In 1975, a new language, called SL5 (''SNOBOL Language 5''), was developed to allow experimentation with a wider variety of programming-language constructs, especially a sophisticated procedure mechanism (Griswold and Hanson 1977; Hanson and Griswold 1978). SL5 extended earlier work in pattern matching, but pattern matching remained essentially a sublanguage with its own control structures, separate from the rest of the language.

The inspiration for Icon came in 1976 with a realization that the control structures that were so useful in pattern matching could be integrated with conventional computational control structures to yield a more coherent and powerful programming language.

The first implementation of Icon (Griswold and Hanson 1979) was written in Ratfor, a preprocessor for Fortran that supports structured programming features (Kernighan 1975). Portability was a central concern in this implementation. The implementation of Icon described in this book is a successor to that first implementation. It borrows much from earlier implementations of SNOBOL4, SL5, and the Ratfor implementation of Icon. As such, it is a distillation and refinement of implementation techniques that have been developed over a period of more than twenty years.

CHAPTER 2

# Icon Language Overview

PERSPECTIVE: The implementor of a programming language needs a considerably different understanding of the language from the persons who are going to use it. An implementor must have a deep understanding of the relationships that exist among various aspects of the language and a precise knowledge of what each operation means. Special cases and details often are of particular importance to the implementor. Users of a language, on the other hand, must know how to use features to accomplish desired results. They often can get by with a superficial knowledge of the language, and they often can use it effectively even if some aspects of the language are misunderstood. Users can ignore parts of the language that they do not need. Idiosyncrasies that plague the implementor may never be encountered by users. Conversely, a detail the implementor overlooks may bedevil users. Furthermore, the implementor may also need to anticipate ways in which users may apply some language features in inefficient and inappropriate ways.

This is a book about the implementation of Version 6 of Icon. The description that follows concentrates on aspects of the language that are needed to understand its implementation. Where there are several similar operations or where the operations are similar to those in well-known programming languages, only representative cases or highlights are given. A complete description of Icon for the user is contained in Griswold and Griswold (1983) and Griswold, Mitchell, and O'Bagy (1986).

Icon is an unusual programming language, and its unusual features are what make its implementation challenging and interesting. The interesting features are semantic, not syntactic; they are part of what the language can do, not part of its appearance. Syntactic matters and the way they are handled in the implementation are of little interest here. The description that follows indicates syntax mostly by example.

This chapter is divided into two major parts. The first part describes the essential aspects of Icon. The second part discusses those aspects of Icon that present the most difficult implementation problems and that affect the nature of the implementation in the most significant ways.

## 2.1 THE ICON PROGRAMMING LANGUAGE

Icon is conventional in many respects. It is an imperative, procedural language with variables, operations, functions, and conventional data types. Its novel aspects lie in its emphasis on the manipulation of strings and structures and in its expression-evaluation mechanism. While much of the execution of an Icon program has an imperative flavor, there also are aspects of logic programming.

There are no type declarations in Icon. Instead, variables can have any type of value. Structures may be heterogeneous, with different elements having values of different types. Type checking is performed during program execution, and automatic type conversion is provided. Several operations are polymorphic, performing different operations depending on the types of their arguments.

Strings and structures are created during program execution, instead of being declared and allocated during compilation. Structures have pointer semantics; a structure value is a pointer to an object. Storage management is automatic. Memory is allocated as required, and garbage collection is performed when necessary. Except for the practical considerations of computer architecture and the amount of available memory, there are no limitations on the sizes of objects.

An Icon program consists of a series of declarations for procedures, records, and global identifiers. Icon has no block structure. Scoping is static: identifiers either are global or are local to procedures.

Icon is an expression-based language with a reserved-word syntax. It resembles C in appearance, for example (Kernighan and Ritchie 1978).

### 2.1.1 Data Types

Icon has many types of data—including several that are not found in most programming languages. In addition to the usual integers and real (floating-point) numbers, there are strings of characters and sets of characters (csets). There is no character data type, and strings of characters are data objects in their own right, not arrays of characters.

There are four structure data types that comprise aggregates of values: lists, sets, tables, and records. Lists provide positional access (like vectors), but they also can be manipulated like stacks and queues. Sets are unordered collections of values on which the usual set operations can be performed. Tables can be subscripted with any kind of value and provide an associative-access mechanism. Records are aggregates of values that can be referenced by name. Record types also add to the built-in type repertoire of Icon.

The null value serves a special purpose; all variables have the null value initially. The null value is illegal in most computational contexts, but it serves to indicate default values in a number of situations. The keyword &null produces the null value.

A source-language file is a data value that provides an interface between the program and a data file in the environment in which the program executes.

Procedures also are data values—''first-class data objects'' in LISP parlance. Procedures can be assigned to variables, transmitted to and returned from functions, and so forth. There is no method for creating procedures during program execution, however.

Finally, there is a co-expression data type. Co-expressions are the expression-level analog of coroutines. The importance of co-expressions is derived from Icon's expression-evaluation mechanism.

Icon has various operations on different types of data. Some operations are polymorphic and accept arguments of different types. For example, type(x) produces a string corresponding to the type of x. Similarly, copy(x) produces a copy of x, regardless of its type. Other operations only apply to certain types. An example is:

```
*x
```

which produces the size of x, where the value of x may be a string, a structure, and so on. Another example is ?x, which produces a randomly selected integer between 1 and x, if x is an integer, but a randomly selected one-character substring of x if x is a string, and so on. In other cases, different operations for similar kinds of computations are syntactically distinguished. For example,

```
i = j
```

compares the numeric values of i and j, while

```
s1 == s2
```

compares the string values of s1 and s2. There is also a general comparison operation that determines whether any two objects are the same:

```
x1 === x2
```

As mentioned previously, any kind of value can be assigned to any variable. For example, x might have an integer value at one time and a string value at another:

```
x := 3
   .
   .
   .
x := "hello"
```

Type checking is performed during program execution. For example, in

```
i := x + 1
```

the value of x is checked to be sure that it is numeric. If it is not numeric, an attempt is made to convert it to a numeric type. If the conversion cannot be performed, program execution is terminated with an error message.

Various conversions are supported. For example, a number always can be converted to a string. Thus,

```
write(*s)
```

automatically converts the integer returned by *s to a string for the purpose of output.

There also are explicit type-conversion functions. For example,

```
s1 := string(*s2)
```

assigns to s1 a string corresponding to the size of s2.

A string can be converted to a number if it has the syntax of a number. Thus,

```
i := i + "20"
```

produces the same result as

```
i := i + 20
```

Augmented assignments are provided for binary operations such as the previous one, where assignment is made to the same variable that appears as the left argument of the operation. Therefore, the previous expression can be written more concisely as

```
i +:= 20
```

Icon also has the concept of a numeric type, which can be either an integer or a real (floating-point) number.

### 2.1.2 Expression Evaluation

In most programming languages—Algol, Pascal, PL/I, and C, for example—the evaluation of an expression always produces exactly one result. In Icon, the evaluation of an expression may produce a single result, it may produce no result at all, or it may produce a sequence of results.

**Success and Failure.** Conventional operations in Icon produce one result, as they do in most programming languages. For example,

```
i + j
```

produces a single result, the sum of the values of i and j. However, a comparison operation, such as

```
i > j
```

produces a result (the value of j) if the value of i is greater than the value of j but does not produce a result if the value of i is not greater than j.

An expression that does not produce a result is said to *fail*, while an expression that produces a result is said to *succeed.* Success and failure are used in several control structures to control program flow. For example,

```
if i > j then write(i) else write(j)
```

writes the maximum of i and j. Note that comparison operations do not produce Boolean values and that Boolean values are not used to drive control structures. Indeed, Icon has no Boolean type.

Generally speaking, an operation that cannot perform a computation does not produce a result, and hence it fails. For example, type-conversion functions fail if the conversion cannot be performed. An example is numeric(x), which converts x to a numeric value if possible, but fails if the conversion cannot be performed. Failure of an expression to produce a result does not indicate an error. Instead, failure indicates that a result does not exist. An example is provided by the function find(s1,s2), which produces the position of s1 as a substring of s2 but fails if s1 does not occur in s2. For example,

```
find("it","They sit like bumps on a log.")
```

produces the value 7 (positions in strings are counted starting at 1). However,

```
find("at","They sit like bumps on a log.")
```

does not produce a result. Similarly, read(f) produces the next line from the file f but fails when the end of the file is reached.

Failure provides a natural way to control loops. For example,

```
while line := read(f) do
   write(line)
```

writes the lines from the file f until an end of file causes read to fail, which terminates the loop.

Another use of success and failure is illustrated by the operation

*expr*

which fails if *expr* is null-valued but produces the result of *expr* otherwise. Since variables have the null value initially, this operation may be used to determine whether a value has been assigned to an identifier, as in

```
if \x then write(x) else write("x is null")
```

If an expression that is enclosed in another expression does not produce a result, there is no value for the enclosing expression, it cannot perform a computation, and it also produces no result. For example, in

```
write(find("at", "They sit like bumps on a log."))
```

the evaluation of find fails, there is no argument for write, and no value is written. Similarly, in

```
i := find("at", "They sit like bumps on a log.")
```

the assignment is not performed and the value of i is not changed.

This ''inheritance'' of failure allows computations to be expressed concisely. For example,

```
while write(read(f))
```

writes the lines from the file f just as the previous loop (the do clause in while-do is optional).

The expression

not *expr*

inverts success and failure. It fails if *expr* succeeds, but it succeeds, producing the null value, if *expr* fails.

Some expressions produce variables, while others only produce values. For example,

```
i + j
```

produces a value, while

```
i := 10
```

produces its left-argument variable. The term *result* is used to refer to a value or a variable. The term *outcome* is used to refer to the consequences of evaluating an expression—either its result or failure.

**Loops.** There are several looping control structures in Icon in addition to while-do. For example,

until $expr_1$ do $expr_2$

evaluates $expr_2$ repeatedly until $expr_1$ succeeds. The control structure

repeat *expr*

simply evaluates *expr* repeatedly, regardless of whether it succeeds or fails.

A loop itself produces no result if it completes, and hence it fails if used in a conditional context. That is, when

while $expr_1$ do $expr_2$

terminates, its outcome is failure. This failure ordinarily goes unnoticed, since loops usually are not used as arguments of other expressions.

The control structure

```
break expr
```

causes the immediate termination of the evaluation of the loop in which it appears, and control is transferred to the point immediately after the loop. The outcome of the loop in this case is the outcome of *expr*. If *expr* is omitted, it defaults to the null value.

An example of the use of break is:

```
while line := read(f) do
   if line == "end" then break
   else write(line)
```

Evaluation of the loop terminates if read fails or if the file f contains a line consisting of "end".

The expression next causes transfer to the beginning of the loop in which it occurs. For example,

```
while line := read(f) do
   if line == "comment" then next
   else write(line)
```

does not write the lines of f that consist of "comment".

The break and next expressions can occur only in loops, and they apply to the innermost loop in which they appear. The argument of break can be a break or next expression, however, so that, for example,

```
break break next
```

breaks out of two levels of loops and transfers control to the beginning of the loop in which they occur.

**Case Expressions.** The case expression provides a way of selecting one of several expressions to evaluate based on the value of a control expression, rather than its success or failure. The case expression has the form

```
case expr of {
   case clauses
        .
        .
        .
   }
```

The value of *expr* is used to select one of the case clauses. A case clause has the form

$expr_1$ : $expr_2$

where the value of *expr* is compared to the value of $expr_1$, and $expr_2$ is evaluated if the comparison succeeds. There is also a default case clause, which has the form

```
default: expr3
```

If no other case clause is selected, *expr3* in the default clause is evaluated. An example is

```
case line := read(f) of {
   "end" :          write("*** end ***")
   "comment" :      write("*** comment ***")
   default :        write(line)
   }
end
```

If the evaluation of the control clause fails, as for an end of file in this example, the entire case expression fails. Otherwise, the outcome of the case expression is the outcome of evaluating the selected expression.

**Generators.** As mentioned previously, an expression may produce a sequence of results. This occurs in situations in which there is more than one possible result of a computation. An example is

```
find("e","They sit like bumps on a log.")
```

in which both 3 and 13 are possible results.

While most programming languages produce only the first result in such a situation, in Icon the two results are produced one after another if the surrounding context requires both of them. Such expressions are called *generators* to emphasize their capability of producing more than one result.

There are two contexts in which a generator can produce more than one result: *iteration* and *goal-directed evaluation*.

Iteration is designated by the control structure

```
every expr1 do expr2
```

in which *expr1* is repeatedly *resumed* to produce its results. For each such result, *expr2* is evaluated. For example,

```
every i := find("e","They sit like bumps on a log.") do
   write(i)
```

writes 3 and 13.

If the argument of an expression is a generator, the results produced by the generator are provided to the enclosing expression—the sequence of results is inherited. Consequently, the previous expression can be written more compactly as

```
every write(find("e","They sit like bumps on a log."))
```

Unlike iteration, which resumes a generator repeatedly to produce all its results, goal-directed evaluation resumes a generator only as necessary, in an attempt to cause an enclosing expression to succeed. While iteration is explicit and occurs only where specified, goal-directed evaluation is implicit and is an inherent aspect of Icon's expression-evaluation mechanism.

Goal-directed evaluation is illustrated by

```
if find("e","They sit like bumps on a log.") > 10
then write("found")
```

The first result produced by find is 3, and the comparison operation fails. Because of goal-directed evaluation, find is automatically resumed to produce another value. Since this value, 13, is greater than 10, the comparison succeeds, and found is written. On the other hand, in

```
if find("e","They sit like bumps on a log.") > 20
then write("found")
```

the comparison fails for 3 and 13. When find is resumed again, it does not produce another result, the control clause of if-then fails, and nothing is written.

There are several expressions in Icon that are generators, including string analysis functions that are similar in nature to find. Another generator is

```
i to j by k
```

which generates the integers from i to j by increments of k. If the by clause is omitted, the increment defaults to one.

The operation !x is polymorphic, generating the elements of x for various types. The meaning of "element" depends on the type of x. If x is a string, !x generates the one-character substrings of x, so that !"hello" generates "h", "e", "l", "l", and "o". If x is a file, !x generates the lines of the file, and so on.

**Generative Control Structures.** There are several control structures related to generators. The *alternation* control structure,

$expr_1$ | $expr_2$

generates the results of $expr_1$ followed by the results of $expr_2$. For example,

```
every write("hello" | "howdy")
```

writes two lines, hello and howdy.

Since alternation succeeds if either of its arguments succeeds, it can be used to produce the effect of logical disjunction. An example is

if (i > j) | (j > k) then *expr*

which evaluates *expr* if i is greater than j or if j is greater than k.

Logical conjunction follows as a natural consequence of goal-directed evaluation. The operation

$expr_1$ & $expr_2$

is similar to other binary operations, such as $expr_1 + expr_2$, except that it performs no computation. Instead, it produces the result of $expr_2$, provided that both $expr_1$ and $expr_2$ succeed. For example,

if (i > j) & (j > k) then *expr*

evaluates *expr* only if i is greater than j and j is greater than k.

Repeated alternation,

|*expr*

generates the results of *expr* repeatedly and is roughly equivalent to

*expr* | *expr* | *expr* | ...

However, if *expr* fails, the repeated alternation control structure stops generating results. For example,

```
|read(f)
```

generates the lines from the file f (one line for each repetition of the alternation) but stops when read(f) fails.

Note that a generator may be capable of producing an infinite number of results. For example,

```
|(1 to 3)
```

can produce 1, 2, 3, 1, 2, 3, 1, 2, 3, ... . However, only as many results as are required by context are actually produced. Thus,

```
i := |(1 to 3)
```

only assigns the value 1 to i, since there is no context to cause the repeated alternation control structure to be resumed for a second result.

The *limitation* control structure

$expr_1$ \ $expr_2$

limits $expr_1$ to at most $expr_2$ results. Consequently,

```
|(1 to 3) \ 5
```

is only capable of producing 1, 2, 3, 1, 2.

**The Order of Evaluation.** With the exception of the limitation control structure, argument evaluation in Icon is strictly left-to-right. The resumption of expressions to produce additional results is in last-in, first-out order. The result is "cross-product" generation of results in expressions that contain several generators. For example,

```
every write((10 to 30 by 10) + (1 to 3))
```

writes 11, 12, 13, 21, 22, 23, 31, 32, 33.

**Control Backtracking.** Goal-directed evaluation results in control backtracking to obtain additional results from expressions that have previously produced results, as in

```
if find("e","They sit like bumps on a log.") > 10
then write("found")
```

Control backtracking is limited by a number of syntactic constructions. For example, in

if $expr_1$ then $expr_2$ else $expr_3$

if $expr_1$ succeeds, but $expr_2$ fails, $expr_1$ is not resumed for another result. (If it were, the semantics of this control structure would not correspond to what "if-then-else" suggests.) Such an expression is called a *bounded expression.* The control clauses of loops also are bounded, as are the expressions within compound expressions:

{ $expr_1$; $expr_2$; $expr_3$; ...; $expr_n$ }

These expressions are evaluated in sequence, but once the evaluation of one is complete (whether it succeeds or fails), and the evaluation of another begins, there is no possibility of backtracking into the preceding one. The last expression in a compound expression is not bounded, however.

Except in such specific situations, expressions are not bounded. For example, in

if $expr_1$ then $expr_2$ else $expr_3$

neither $expr_2$ nor $expr_3$ is bounded. Since Icon control structures are expressions that may return results, it is possible to write expressions such as

```
every write(if i > j then j to i else i to j)
```

which writes the integers from i to j in ascending sequence.

**Data Backtracking.** While control backtracking is a fundamental part of expression evaluation in Icon, data backtracking is not performed except in a few specific operations. For example, in

```
(i := 3) & read(f)
```

the value of 3 is assigned to i. Even if read(f) fails, the former value of i is not restored.

There are, however, specific operations in which data backtracking is performed. For example, the *reversible assignment* operation

```
x <- y
```

assigns the value of y to x, but it restores the former value of x if control backtracking into this expression occurs. Thus,

```
(i <- 3) & read(f)
```

assigns 3 to i but restores the previous value of i if read(f) fails.

### 2.1.3 Csets and Strings

Csets are unordered sets of characters, while strings are sequences of characters. There are 256 different characters, the first 128 of which are interpreted as ASCII. The number and interpretation of characters is independent of the architecture of the computer on which Icon is implemented.

**Csets.** Csets are represented literally by surrounding their characters by single quotation marks. For example,

```
vowels := 'aeiouAEIOU'
```

assigns a cset of 10 characters to vowels.

There are several built-in csets that are the values of keywords. These include &lcase, &ucase, and &cset, which contain the lowercase letters, the uppercase letters, and all 256 characters, respectively.

Operations on csets include union, intersection, difference, and complement with respect to &cset. Csets are used in lexical analysis. For example, the function upto(c, s) is analogous to find(s1, s2), except that it generates the positions at which any character of c occurs in s. Thus,

```
upto(vowels, "They sit like bumps on a log.")
```

is capable of producing 3, 7, 11, 13, 16, 21, 24, and 27.

**Strings.** Strings are represented literally by surrounding their characters with double quotation marks instead of single quotation marks. The empty string, which contains no characters, is given by "". The size of a string is given by *s. For example, if

```
command := "Sit still!"
```

then the value of *command is 10. The value of *"" is 0. Space for strings is provided automatically and there is no inherent limit to the size of a string.

There are several operations that construct strings. The principal one is concatenation, denoted by

```
s1 || s2
```

The function repl(s, i) produces the result of concatenating s i times. Thus,

```
write(repl("*!", 3))
```

writes *!*!*!.

Other string construction functions include reverse(s), which produces a string with the characters of s in reverse order, and trim(s, c), which produces a string in which any trailing characters of s that occur in c are omitted. There also are functions for positioning a string in a field of a fixed width. For example, the function left(s1, i, s2) produces a string of length i positioned at the left and padded with copies of s2 as needed.

Substrings are produced by subscripting a string with the beginning and ending positions of the desired substring. Positions in strings are between characters, and the position before the first character of a string is numbered 1. For example,

```
verb := command[1:4]
```

assigns the string "Sit" to verb. Substrings also can be specified by the beginning position and a length, as in

```
verb := command[1+:3]
```

If the length of a substring is 1, only the first position need be given, so that the value of command[2] is "i".

Assignment can be made to a subscripted string to produce a new string. For example,

```
command[1:4] := "Remain"
```

changes the value of command to "Remain still!".

String operations are applicative; no operation on a string in Icon changes the characters in it. The preceding example may appear to contradict this, but in fact

```
command[1:4] := "Remain"
```

is an abbreviation for

```
command := "Remain" || command[5:11]
```

Thus, a new string is constructed and then assigned to command.

Nonpositive values can be used to specify a position with respect to the right end of a string. For example, the value of command[−1] is "!". The value 0 refers to the position after the last character of a string, so that if the value of command is "Sit still!",

```
command[5:0]
```

is equivalent to

```
command[5:11]
```

The subscript positions can be given in either order. Thus,

```
command[11:5]
```

produces the same result as

```
command[5:11]
```

String-analysis functions like find and upto have optional third and fourth arguments that allow their range to be restricted to a particular portion of a string. For example,

```
upto(vowels,"They sit like bumps on a log.",10,20)
```

only produces positions of vowels between positions 10 and 20 of its second argument: 11, 13, and 16. If these arguments are omitted, they default to 1 and 0, so that the entire string is included in the analysis.

**Mapping.** One of the more interesting string-valued functions in Icon is map(s1,s2,s3). This function produces a string obtained from a character substitution on s1. Each character of s1 that occurs in s2 is replaced by the corresponding character in s3. For example,

```
write(map("Remain still!","aeiou","*****"))
```

writes R*m**n St*ll!. Characters in s1 that do not appear in s2 are unchanged, as this example shows. If a character occurs more than once in s2, its second correspondence in s3 applies. Consequently,

```
s2 := &lcase || &ucase || "aeiou"
s3 := repl("l",26) || repl("u",26) || "*****"
write(map("Remain still!",s2,s3))
```

writes u*l**l ll*ll!.

### 2.1.4 String Scanning

String scanning is a high-level facility for string analysis that suppresses the computational details associated with the explicit location of positions and substring specifications. In string scanning, a subject serves as a focus of attention. A position in this subject is maintained automatically.

A string-scanning expression has the form

$expr_1$ ? $expr_2$

in which the evaluation of $expr_1$ provides the subject. The position in the subject is 1 initially. The expression $expr_2$ is then evaluated in the context of this subject and position.

Although $expr_2$ can contain any operation, two *matching functions* are useful in analyzing the subject:

| | |
|---|---|
| tab(i) | set the position in the subject to i |
| move(i) | increment the position in the subject by i |

Both of these functions return the substring of the subject between the old and new positions. If the position is out of the range of the subject, the matching function fails and the position is not changed. The position can be increased or decreased. Nonpositive values can be used to refer to positions relative to the end of the subject. Thus, tab(0) moves the position to the end of the subject, matching the remainder of the subject.

An example of string scanning is

```
line ? while write(move(2))
```

which writes successive two-character substrings of line, stopping when there are not two characters remaining.

In string scanning, the trailing arguments of string analysis functions such as find and upto are omitted; the functions apply to the subject at the current position. Therefore, such functions can be used to provide arguments for matching functions. An example is

```
line ? write(tab(find("::=")))
```

which writes the initial portion of line up to an occurrence of the string "::=".

If a matching function is resumed, it restores the position in the subject to the value that it had before the matching function was evaluated. For example, suppose that line contains the substring "::=". Then

```
line ? ((tab(find("::=") + 3)) & write(move(10)) | write(tab(0)))
```

writes the 10 characters after "::=", provided there are 10 more characters. However, if there are not 10 characters remaining, move(10) fails and tab(find("::=")) is resumed. It restores the position to the beginning of the subject, and the alternative, tab(0), matches the entire subject, which is written.

Data backtracking of the position in the subject is important, since it allows matches to be performed with the assurance that any previous alternatives that failed to match left the position where it was before they were evaluated.

The subject and position are directly accessible as the values of the keywords &subject and &pos, respectively. For example,

```
&subject := "Hello"
```

assigns the string "Hello" to the subject. Whenever a value is assigned to the subject, &pos is set to 1 automatically.

The values of &subject and &pos are saved at the beginning of a string-scanning expression and are restored when it completes. Consequently, scanning expressions can be nested.

### 2.1.5 Lists

A list is a linear aggregate of values (''elements''). For example,

```
cities := ["Portland","Toledo","Tampa"]
```

assigns a list of three strings to cities. Lists can be heterogeneous, as in

```
language := ["Icon",1978,"The University of Arizona"]
```

An empty list, containing no elements, is produced by []. The function

```
list(i, x)
```

produces a list of i elements, each of which has the value of x. The size operation *x also applies to lists. The value of *cities is 3, for example.

An element of a list is referenced by a subscripting expression that has the same form as the one for strings. For example,

```
cities[3] := "Miami"
```

changes the value of cities to

```
["Portland","Toledo","Miami"]
```

The function sort(a) produces a sorted copy of a. For example, sort(cities) produces

```
["Miami","Portland","Toledo"]
```

List operations, unlike string operations, are not applicative. While assignment to a substring is an abbreviation for concatenation, assignment to a subscripted list changes the value of the subscripted element.

A list value is a pointer to a structure that contains the elements of the list. Assignment of a list value copies this pointer, but it does not copy the structure. Consequently, in

```
states := ["Nevada","Texas","Maine","Georgia"]
slist := states
```

both states and slist point to the *same* structure. Because of this,

```
states[2] := "Arkansas"
```

changes the second element of slist as well as the second element of states.

The elements of a list may be of any type, including lists, as in

```
tree := ["a", ["b", ["c"], ["d"]]]
```

which can be depicted as

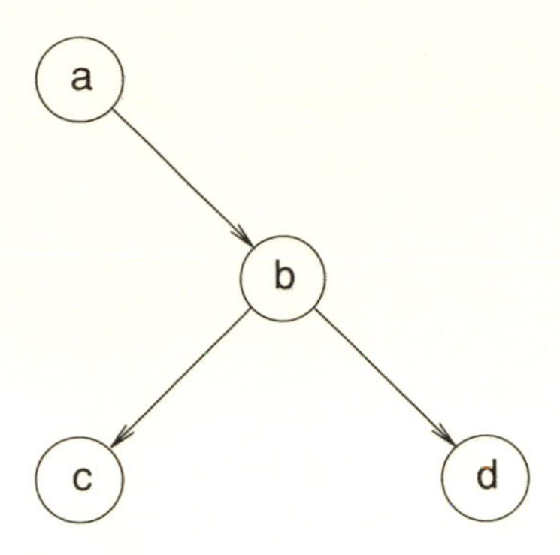


Structures also can be used to represent loops, as in

```
graph := ["a", ""]
graph[2] := graph
```

which can be depicted as

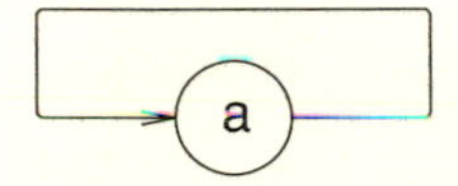


Lists are not fixed in size. Elements can be added to them or removed from them at their ends by queue and stack functions.

The function put(a, x) adds the value of x to the right end of the list a, increasing its size by one. Similarly, push(a, x) adds the value of x to the left end of a. For example,

```
lines := []
while put(lines, read(f))
```

constructs a list of the lines from the file f. Conversely,

```
lines := []
while push(lines, read(f))
```

constructs a list of lines in reverse order.

The functions pop(a) and get(a) are the same. They both remove an element from the left end of a and return it as the value of the function call, but they fail if a is empty. Consequently,

```
lines := []
while push(lines, read(f))
while write(pop(lines))
```

writes out the lines of f in reverse order. The function pull(a) is similar, but it removes an element from the right end of a.

Other operations on lists include concatenation, which is denoted by

```
a1 ||| a2
```

where a1 and a2 are lists. There is no automatic conversion of other types to lists.

List sectioning is denoted by

```
a[i:j]
```

The result is a *new* list containing values i through j of a.

There is no inherent limit to the size of a list, either when it is originally created or as a result of adding elements to it.

### 2.1.6 Sets

A set is an unordered collection of values. Unlike csets, which contain only characters, sets are collections of Icon values that can be of any type. A set is constructed from a list by set(a). For example,

```
states := set(["Virginia", "Rhode Island", "Kansas", "Illinois"])
```

assigns a set of four elements to states.

The operation

```
member(s, x)
```

succeeds if the value of x is a member of s but fails otherwise. The operation

```
insert(s, x)
```

adds the value of x to s if it is not already a member of s, while

```
delete(s, x)
```

deletes the value of x from s. The operations of union, intersection, and difference for sets also are provided.

Like other structures, sets can be heterogeneous. A set can even be a member of itself, as in

```
insert(s, s)
```

There is no contradiction here, since a set value is a pointer to the structure for the set.

### 2.1.7 Tables

A table is a set of pairs of values. Tables provide an associative lookup mechanism as contrasted with positional references to lists. They can be subscripted with an *entry value* to which a value can be assigned to make up a pair, called a table element.

A table is created by

```
table(x)
```

Tables are empty initially. The value of x is an assigned default value that is produced if the table is subscripted with an entry value to which no value has been assigned (that is, for an element that is not in the table). For example,

```
states := table(0)
```

assigns to states a table with a default value of 0. An element can be added to states by an assignment such as

```
states["Oregon"] := 1
```

which adds a table element for "Oregon" with the value 1 to states. On the other hand,

```
write(states["Utah"])
```

writes 0, the default value, if there is no element in the table for "Utah".

Tables can be heterogeneous and have a mixture of types for entry and assigned values. Tables grow automatically in size as new elements are added and there is no inherent limit on the size of a table.

### 2.1.8 Records

A record is an aggregate of values that is referenced by named fields. Each record type has a separate name. A record type and the names of its fields are given in a declaration. For example,

```
record rational(numerator, denominator)
```

declares a record of type rational with two fields: numerator and denominator.

An instance of a record is created by calling a record-constructor function corresponding to the form of the declaration for the record type. Thus,

```
r := rational(3, 5)
```

assigns to r a record of type rational with a numerator field of 3 and a denominator field of 5. Fields are referenced by name, as in

```
write(r.numerator)
```

which writes 3. Fields can also be referred to by position; r[1] is equivalent to r.numerator.

There is no inherent limit to the number of different record types. The same field names can be given for different record types, and such fields need not be in the same position for all such record types.

### 2.1.9 Input and Output

Input and output in Icon are sequential and comparatively simple. The standard input, standard output, and standard error output files are the values of &input, &output, and &errout, respectively. The function

```
open(s1, s2)
```

opens the file whose name is s1 according to options given by s2 and produces a value of type file. Typical options are "r" for opening for reading and "w" for opening for writing. The default is "r". For example,

```
log := open("grade.log", "w")
```

assigns a value of type file to log, corresponding to the data file grade.log, which is opened for writing. The function open fails if the specified file cannot be opened according to the options given. The function close(f) closes the file f.

The function read(f) reads a line from the file f but fails if an end of file is encountered. The default is standard input if f is omitted.

The result of

```
write(x1, x2, ..., xn)
```

depends on the types of x1, x2, ..., xn. Strings and types convertible to strings are written, but if one of the arguments is a file, subsequent strings are written to that file. The default file is standard output. Thus,

```
write(s1, s2)
```

writes the concatenation of s1 and s2 to standard output, but

```
write(log, s)
```

writes s to the file grade.log. In any event, write returns the string value of the last argument written.

The function

```
stop(x1, x2, ..., xn)
```

produces the same output as write, but it then terminates program execution.

### 2.1.10 Procedures

**Procedure Declarations.** The executable portions of an Icon program are contained in procedure declarations. Program execution begins with a call of the procedure main.

An example of a procedure declaration is:

```
procedure maxstr(slist)
   local max, value
   max := 0
   every value := *!slist do
      if value > max then max := value
   return max
end
```

This procedure computes the longest string in a list of strings. The formal parameter slist and the identifiers max and value are local to calls of the procedure maxstr. Storage for them is allocated when maxstr is called and deallocated when maxstr returns.

A procedure call has the same form as a function call. For example,

```
lines := []
while put(lines, read(f))
write(maxstr(lines))
```

writes the length of the longest line in the file f.

A procedure call may fail to produce a result in the same way that a built-in operation can fail. This is indicated by fail in the procedure body in place of return. For example, the following procedure returns the length of the longest string in slist but fails if that length is less than limit:

```
procedure maxstr(slist, limit)
   local max, value
   max := 0
   every value := *!slist do
      if value > max then max := value
   if max < limit then fail else return max
end
```

Flowing off the end of a procedure body without an explicit return is equivalent to fail.

A procedure declaration may have static identifiers that are known only to calls of that procedure but whose values are not destroyed when a call returns. A procedure declaration also may have an initial clause whose expression is evaluated only the first time the procedure is called. The use of a static identifier and an initial clause is illustrated by the following procedure, which returns the longest of all the strings in the lists it has processed:

```
procedure maxstrall(slist)
   local value
   static max
   initial max := 0
   every value := *!slist do
      if value > max then max := value
   return max
end
```

**Procedures and Functions.** Procedures and functions are used in the same way. Their names have global scope. Other identifiers can be declared to have global scope, as in

```
global count
```

Such global declarations are on a par with procedure declarations and cannot occur within procedure declarations.

A call such as

```
write(maxstr(lines))
```

applies the *value* of the identifier maxstr to lines and applies the *value* of the identifier write to the result. There is nothing fixed about the values of such identifiers. In this case, the initial value of maxstr is a procedure, as a consequence of the procedure declaration for it. Similarly, the initial value of write is a function. These values can be assigned to other variables, as in

```
print := write
      .
      .
      .
print(maxstr(lines))
```

in which the function that is the initial value of write is assigned to print.

Similarly, nothing prevents an assignment to an identifier whose initial value is a procedure. Consequently,

```
write := 3
```

assigns an integer to write, replacing its initial function value.

Although it is typical to call a procedure by using an identifier that has the procedure value, the procedure used in a call can be computed. The general form of a call is

$$expr_0(expr_1, expr_2, \ldots, expr_n)$$

where the value of $expr_0$ is applied to the arguments resulting from the evaluation of $expr_1$, $expr_2$, ..., $expr_n$. For example,

(proclist[i])($expr_1$, $expr_2$, ..., $expr_n$)

applies the procedure that is the ith element of proclist.

Procedures may be called recursively. The recursive nature of a call depends on the fact that procedure names are global. The "Fibonacci strings" provide an example:

```
procedure fibstr(i)
   if i = 1 then return "a"
   else if i = 2 then return "b"
   else return fibstr(i - 1) || fibstr(i - 2)
end
```

An identifier that is not declared in a procedure and is not global defaults to local. Thus, local declarations can be omitted, as in

```
procedure maxstr(slist)
   max := 0
   every value := *!slist do
      if value > max then max := value
   return max
end
```

**Procedures as Generators.** In addition to returning and failing, a procedure can also suspend. In this case, the values of its arguments and local identifiers are not destroyed, and the call can be resumed to produce another result in the same way a built-in generator can be resumed. An example of such a generator is

```
procedure intseq(i)
   repeat {
      suspend i
      i +:= 1
      }
end
```

A call intseq(10), for example, is capable of generating the infinite sequence of integers 10, 11, 12, ... . For example,

```
every f(intseq(10) \ 5)
```

calls f(10), f(11), f(12), f(13), and f(14).

If the argument of suspend is a generator, the generator is resumed when the call is resumed and the call suspends again with the result it produces. A generator of the Fibonacci strings provides an example:

```
procedure fibstrseq()
   local s1, s2, s3
   s1 := "a"
   s2 := "b"
   suspend (s1 | s2)
   repeat {
      suspend s3 := s1 || s2
      s1 := s2
      s2 := s3
      }
end
```

When this procedure is called, the first suspend expression produces the value of s1, "a". If the call of fibstrseq is resumed, the argument of suspend is resumed and produces the value of s2, "b". If the call is resumed again, there is no further result for the first suspend, and evaluation continues to the repeat loop.

Repeated alternation often is useful in supplying an endless number of alternatives. For example, the procedure intseq(i) can be rewritten as

```
procedure intseq(i)
   suspend i | (i +:= |1)
end
```

Note that |1 is used to provide an endless sequence of increments.

**Argument Transmission.** Omitted arguments in a procedure or function call (including trailing ones) default to the null value. Extra arguments are evaluated, but their values are discarded.

Some functions, such as write, may be called with an arbitrary number of arguments. All arguments to procedures and functions are passed by value. If the evaluation of an argument expression fails, the procedure or function is not called. This applies to extra arguments. Arguments are not dereferenced until all of them have been evaluated. Dereferencing cannot fail. Since no argument is dereferenced until all argument expressions are evaluated, expressions with side effects can produce unexpected results. Thus, in

```
write(s, s := "hello")
```

the value written is hellohello, regardless of the value of s before the evaluation of the second argument of write.

**Dereferencing in Return Expressions.** The result returned from a procedure call is dereferenced unless it is a global identifier, a static identifier, a subscripted structure, or a subscripted string-valued global identifier.

In these exceptional cases, the variable is returned and assignment can be made to the procedure call. An example is

```
procedure maxel(a, i, j)
   if i > j then return a[i]
   else return a[j]
end
```

Here a list element, depending on the values of i and j, is returned. An assignment can be made to it, as in

```
maxel(lines, i, j) := "end"
```

which assigns "end" to lines[i] or lines[j], depending on the values of i and j.

**Mutual Evaluation.** In a call expression, the value of $expr_0$ can be an integer i as well as a procedure. In this case, called *mutual evaluation*, the result of the ith argument is produced. For example,

```
i := 1(find(s1, line1), find(s2, line2))
```

assigns to i the position of s1 in line1, provided s1 occurs in line1 and that s2 occurs in line2. If either call of find fails, the expression fails and no assignment is made.

The selection integer in mutual evaluation can be negative, in which case it is interpreted relative to the end of the argument list. Consequently,

(−1)($expr_1$, $expr_2$, ..., $expr_n$)

produces the result of $expr_n$ and is equivalent to

$expr_1$ & $expr_2$ & ... & $expr_n$

The selection integer can be omitted, in which case it defaults to −1.

### 2.1.11 Co-Expressions

The evaluation of an expression in Icon is limited to the site in the program where it appears. Its results can be produced only at that site as a result of iteration or goal-directed evaluation. For example, the results generated by intseq(i) described in Sec. 2.1.10 can only be produced where it is called, as in

```
every f(intseq(10) \ 5)
```

It is often useful, however, to be able to produce the results of a generator at various places in the program as the need for them arises. Co-expressions provide this facility by giving a context for the evaluation of an expression that is maintained in a data structure. Co-expressions can be *activated* to produce the results of a generator on demand, at any time and place in the program.

A co-expression is constructed by

```
create expr
```

The expression *expr* is not evaluated at this time. Instead, an object is produced through which *expr* can be resumed at a later time. For example,

```
label := create ("L" || (1 to 100) || ":")
```

assigns to label a co-expression for the expression

```
"L" || (1 to 100) || ":"
```

The operation @label activates this co-expression, which corresponds to resuming its expression. For example,

```
write(@label)
write("    tstl    count")
write(@label)
```

writes

```
L1:
    tstl    count
L2:
```

If the resumption of the expression in a co-expression does not produce a result, the co-expression activation fails. For example, after @label has been evaluated 100 times, subsequent evaluations of @label fail. The number of results that a co-expression e has produced is given by *e.

The general form of the activation expression is

$expr_1$ @ $expr_2$

which activates $expr_2$ and transmits the result of $expr_1$ to it. This form of activation can be used to return a result to the co-expression that activated the current one.

A co-expression is a value like any other value in Icon and can be passed as an argument to a procedure, returned from a procedure, and so forth. A co-expression can survive the call of the procedure in which it is created.

If the argument of a create expression contains identifiers that are local to the procedure in which the create occurs, copies of these local identifiers are included in the co-expression with the values they have at the time the create expression is evaluated. These copied identifiers subsequently are independent of the local identifiers in the procedure. Consider, for example,

```
procedure labgen(tag)
   local i, j
      :
   i := 10
   j := 20
   e := create (tag || (i to j) || ":")
      :
   i := j
   if i > 15 then return e
      :
end
```

The expression

```
labels := labgen("X")
```

assigns to labels a co-expression that is equivalent to evaluating

```
create ("X" || (10 to 20) || ":")
```

The fact that i is changed after the co-expression was assigned to e, but before e returns, does not affect the co-expression, since it contains copies of i and j at the time it was created. Subsequent changes to the values of i or j do not affect the co-expression.

A copy of a co-expression e is produced by the *refresh* operation, ^e. When a refreshed copy of a co-expression is made, its expression is reset to its initial state, and the values of any local identifiers in it are reset to the values they had when the co-expression was created. For example,

```
newlabels := ^labels
```

assigns to newlabels a co-expression that is capable of producing the same results as labels, regardless of whether or not labels has been activated.

The value of the keyword &main is the co-expression for the call of main that initiates program execution.

### 2.1.12 Diagnostic Facilities

**String Images.** The function type(x) only produces the string name of the type of x, but the function image(x) produces a string that shows the value of x. For strings and csets, the value is shown with surrounding quotation marks in the fashion of program literals. For example,

```
write(image("Hi there!"))
```

writes "Hi there!", while

```
write(image('aeiou'))
```

writes 'aeiou'.

For structures, the type name and size are given. For example,

```
write(image([]))
```

writes list(0).

Various forms are used for other types of data, using type names where necessary so that different types of values are distinguishable.

**Tracing.** If the value of the keyword &trace is nonzero, a trace message is produced whenever a procedure is called, returns, fails, suspends, or is resumed. Trace messages are written to standard error output. The value of &trace is decremented for every trace message. Tracing stops if the value of &trace becomes zero, which is its initial value. Suppose that the following program is contained in the file fibstr.icn:

```
procedure main()
   &trace := -1
   fibstr(3)
end

procedure fibstr(i)
   if i = 1 then return "a"
   else if i = 2 then return "b"
   else return fibstr(i - 1) || fibstr(i - 2)
end
```

The trace output of this program is

```
fibstr.icn: 3      | fibstr(3)
fibstr.icn: 9      | | fibstr(2)
fibstr.icn: 8      | | fibstr returned "b"
fibstr.icn: 9      | | fibstr(1)
fibstr.icn: 7      | | fibstr returned "a"
fibstr.icn: 9      | fibstr returned "ba"
fibstr.icn: 4      main failed
```

In addition to the indentation corresponding to the level of procedure call, the value of the keyword &level also is the current level of call.

**Displaying Identifier Values.** The function display(i, f) writes a list of all identifiers and their values for i levels of procedure calls, starting at the current level. If i is omitted, the default is &level, while if f is omitted, the list is written to standard error output. The format of the listing produced by display is illustrated by the following program:

```
procedure main()
   log := open("grade.log", "w")
   while write(log, check(read()))
end

procedure check(value)
   static count
   initial count := 0
   if numeric(value) then {
      count +:= 1
      return value
      }
   else {
     display()
     stop("nonnumeric value")
     }
end
```

Suppose that the tenth line of input is the nonnumeric string "3.a". Then the output of display is

```
check local identifiers:
   value = "3.a"
   count = 9
main local identifiers:
   log = file(grade.log)
global identifiers:
   main = procedure main
   check = procedure check
   open = function open
   write = function write
   read = function read
   numeric = function numeric
   display = function display
   stop = function stop
```

**Error Messages.** If an error is encountered during program execution, a message is written to standard error output and execution is terminated. For example, if the tenth line of a program contained in the file check.icn is

```
i +:= "x"
```

evaluation of this expression produces the error message

```
Run–time error 102 at line 10 in check.icn
numeric expected
offending value: "x"
```

## 2.2 LANGUAGE FEATURES AND THE IMPLEMENTATION

Even a cursory consideration of Icon reveals that some of its features present implementation problems and require approaches that are different from ones used in more conventional languages. In the case of a language of the size and complexity of Icon, it is important to place different aspects of the implementation in perspective and to identify specific problems.

**Values and Variables.** The absence of type declarations in Icon has far-reaching implications. Since any variable may have a value of any type and the type may change from time to time during program execution, there must be a way of representing values uniformly. This is a significant challenge in a language with a wide variety of types ranging from integers to co-expressions. Heterogeneous structures follow as a natural consequence of the lack of type declarations.

In one sense, the absence of type declarations simplifies the implementation: there is not much that can be done about types during program translation (compilation), and some of the work that is normally performed by conventional compilers can be avoided. The problems do not go away, however—they just move to another part of the implementation, since run-time type checking is required. Automatic type conversion according to context goes hand-in-hand with type checking.

**Storage Management.** Since strings and structures are created during program execution, rather than being declared, the space for them must be allocated as needed at run time. This implies, in turn, some mechanism for reclaiming space that has been allocated but which is no longer needed—''garbage collection.'' These issues are complicated by the diversity of types and sizes of objects, the lack of any inherent size limitations, and the possibility of pointer loops in circular structures.

**Strings.** Independent of storage-management considerations, strings require special attention in the implementation. The emphasis of Icon is on string processing, and it is necessary to be able to process large amounts of string data efficiently. Strings may be very long and many operations produce substrings of other strings. The repertoire of string analysis and string synthesis functions is large. All this adds up to the need for a well-designed and coherent mechanism for handling strings.

**Structures.** Icon's unusual structures, with sophisticated access mechanisms, also pose problems. In particular, structures that can change in size and can grow without limit require different implementation approaches than static structures of fixed size and organization.

The flexibility of positional, stack, and queue access mechanisms for lists requires compromises to balance efficient access for different uses. Sets of values with arbitrary types, combined with a range of set operations, pose nontrivial implementation problems. Tables are similar to sets, but require additional attention because of the implicit way that elements are added.

**Procedures and Functions.** Since procedures and functions are values, they must be represented as data objects. More significantly, the meaning of a function call cannot, in general, be determined when a program is translated. The expression write(s) may write a string or it may do something else, depending on whether or not write still has its initial value. Such meanings must, instead, be determined at run time.

**Polymorphic Operations.** Although the meanings of operations cannot be changed during program execution in the way that the meanings of calls can, several operations perform different computations depending on the types of their operands. Thus, x[i] may subscript a string, a list, or a table.

The meanings of some operations also depend on whether they occur in an assignment or a dereferencing context. For example, if s has a string value, assignment to s[i] is an abbreviation for a concatenation followed by an assignment to s, while if s[i] occurs in a context where its value is needed, it is simply a substring operation. Moreover, the context cannot, in general, be determined at translation time.

The way subscripting operations are specified in Icon offers considerable convenience to the programmer at the expense of considerable problems for the implementor.

**Expression Evaluation.** Generators and goal-directed evaluation present obvious implementation problems. There is a large body of knowledge about the implementation of expression evaluation for conventional languages in which expressions always produce a single result, but there is comparatively little knowledge about implementing expressions that produce results in sequence.

While there are languages in which expressions can produce more than one result, this capability is limited to specific contexts, such as pattern matching, or to specific control structures or data objects.

In Icon, generators and goal-directed evaluation are general and pervasive and apply to all evaluation contexts and to all types of data. Consequently, their implementation requires a fresh approach. The mechanism also has to handle the use of failure to drive control structures and must support novel control structures, such as alternation and limitation. Efficiency is a serious concern, since whatever mechanism is used to implement generators is also used in conventional computational situations in which only one result is needed.

**String Scanning.** String scanning is comparatively simple. The subject and position—''state variables''—have to be saved at the beginning of string scanning and restored when it is completed. Actual string analysis and matching follow trivially from generators and goal-directed evaluation.

**Co-Expressions.** Co-expressions, which are only relevant because of the expression-evaluation mechanism of Icon, introduce a whole new set of complexities. Without co-expressions, the results that a generator can produce are limited to its site in the program. Control backtracking is limited syntactically, and its scope can be determined during program translation. With co-expressions, a generator in a state of suspension can be activated at any place and time during program execution.

RETROSPECTIVE: Icon has a number of unusual features that are designed to facilitate programming, and it has an extensive repertoire of string and structure operations. One of Icon's notable characteristics is the freedom from translation-time constraints and the ability to specify and change the meanings of operations at run time. This run-time flexibility is valuable to the programmer, but it places substantial burdens on the implementation—and also makes it interesting.

At the top level, there is the question of how actually to carry out some of the more sophisticated operations. Then there are questions of efficiency, both in execution speed and storage utilization. There are endless possibilities for alternative approaches and refinements.

It is worth noting that many aspects of the implementation are relatively independent of each other and can be approached separately. Operations on strings and structures are largely disjoint and can, except for general considerations of the representation of values and storage management, be treated as independent problems.

The independence of expression evaluation from other implementation considerations is even clearer. Without generators and goal-directed evaluation, Icon would be a fairly conventional high-level string and structure processing language, albeit one with interesting implementation problems. On the other hand, generators and goal-directed evaluation are not dependent in any significant way on string and structure data types. Generators, goal-directed evaluation, and related control structures could just as well be incorporated in a programming language emphasizing numerical computation. The implementation problems related to expression evaluation in the two contexts are largely the same.

While untyped variables and automatic storage management have pervasive effects on the overall implementation of Icon, there are several aspects of Icon that are separable from the rest of the language and its implementation. Any specific data structure, string scanning, or co-expressions could be eliminated from the language without significantly affecting the rest of the implementation. Similarly, new data structures and new access mechanisms could be added without requiring significant modifications to the balance of the implementation.

## EXERCISES

**2.1** What is the outcome of the following expression if the file f contains a line consisting of "end", or if it does not?

```
while line := read(f) do
   if line == "end" then break
   else write(line)
```

**2.2** What does

```
write("hello" | "howdy")
```

write?

**2.3** What is the result of evaluating the following expression?

```
|(1 to 3) > 10
```

**2.4** Explain the rationale for the dereferencing of variables when a procedure call returns.

**2.5** Give an example of a situation in which it cannot be determined until run time whether a string subscripting expression is used in an assignment or a dereferencing context.

CHAPTER 3

# Organization of the Implementation

PERSPECTIVE: Many factors influence the implementation of a programming language. The properties of the language itself, of course, are of paramount importance. Beyond this, goals, resources, and many other factors may affect the nature of an implementation in significant and subtle ways.

In the case of the implementation of Icon described here, several unusual factors deserve mention. To begin with, Icon's origins were in a research project, and its implementation was designed not only to make the language available for use but also to support further language development. The language itself was less well defined and more subject to modification than is usually the case with an implementation. Therefore, flexibility and ease of modification were important implementation goals.

Although the implementation was not a commercial enterprise, neither was it a toy or a system intended only for a few "friendly users." It was designed to be complete, robust, easy to maintain, and sufficiently efficient to be useful for real applications in its problem domain.

Experience with earlier implementations of SNOBOL4, SL5, and the Ratfor implementation of Icon also influenced the implementation that is described here. They provided a repertoire of proven techniques and a philosophy of approach to the implementation of a programming language that has novel features.

The computing environment also played a major role. The implementation started on a PDP-11/70 running under UNIX. The UNIX environment (Ritchie and Thompson 1978), with its extensive range of tools for program development, influenced several aspects of the implementation in a direct way. C (Kernighan and Ritchie 1978) is the natural language for writing such an implementation under UNIX, and its use for the majority of Icon had pervasive effects, which are described throughout this book. Tools, such as the Yacc parser-generator (Johnson 1975), influenced the approach to the translation portion of the implementation.

Since the initial work was done on a PDP-11/70, with a user address space of only 128K bytes (combined instruction and data spaces), the size of the implementation was a significant concern. In particular, while the Ratfor implementation of Icon fit comfortably on computers with large address spaces, such as the DEC-10, CDC Cyber, and IBM 370, this implementation was much too large to fit on a PDP-11/70.

## 3.1 THE ICON VIRTUAL MACHINE

The implementation of Icon is organized around a virtual machine (Newey, Poole, and Waite 1972; Griswold 1977). Virtual machines, sometimes called abstract machines, serve as software design tools for implementations in which the operations of a language do not fit a particular computer architecture or when portability is a consideration and the attributes of several real computer architectures can be abstracted in a single common model. The expectation for most virtual machine models is that a translation will be performed to map the virtual machine operations onto a specific real machine. A virtual machine also provides a basis for developing an operational definition of a programming language in which details can be worked out in concrete terms.

During the design and development phases of an implementation, a virtual machine serves as an idealized model that is free of the details and idiosyncrasies of any real machine. The virtual machine can be designed in such a way that treatment of specific, machine-dependent details can be deferred until it is necessary to translate the implementation of the virtual machine to a real one.

Icon's virtual machine only goes so far. Unlike the SNOBOL4 virtual machine (Griswold 1972), it is incomplete and characterizes only the expression-evaluation mechanism of Icon and computations on Icon data. It does not, *per se*, include a model for the organization of memory. There are many aspects of the Icon run-time system, such as type checking, storage allocation, and garbage collection, that are not represented in the virtual machine. Instead, Icon's virtual machine serves more as a guide and a tool for organizing the implementation than it does as a rigid structure that dominates the implementation.

## 3.2 COMPONENTS OF THE IMPLEMENTATION

There are three major components of the implementation of Icon: a translator, a linker, and a run-time system.

The translator plays the role of a compiler for the Icon virtual machine. It analyzes source programs and converts them to virtual machine instructions. The output of the translator is called *ucode*. Ucode is represented as ASCII text, which is helpful in debugging the implementation.

The linker combines one or more ucode files into a single program for the virtual machine. This allows programs to be written and translated in a number of modules, and it is particularly useful for giving users access to pretranslated libraries of Icon procedures. The output of the linker, called *icode*, is in binary format for compactness and ease of processing by the virtual machine. Ucode and icode instructions are essentially the same, differing mainly in their format.

Translating and linking are done in two phases:

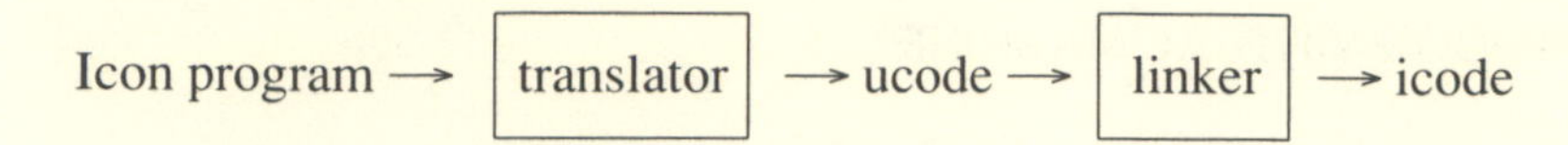


These phases can be performed separately. If only the first phase is performed, the result is ucode, which can be saved and linked at another time.

The run-time system consists of an interpreter for icode and a library of support routines to carry out the various operations that may occur when an Icon program is executed. The interpreter serves, conceptually, as a software realization of the Icon virtual machine. It decodes icode instructions and their operands and carries out the corresponding operations.

It is worth noting that the organization of the Icon system does not depend in any essential way on the use of an interpreter. In fact, in the early versions of this implementation, the linker produced assembly-language code for the target machine. That code then was assembled and loaded with the run-time library. On the surface, the generation of machine code for a specific target machine rather than for a virtual machine corresponds to the conventional compilation approach. However, this is somewhat of an illusion, since the machine code consists largely of calls to run-time library routines corresponding to virtual machine instructions. Execution of machine code in such an implementation therefore differs only slightly from interpretation, in which instruction decoding is done in software rather than in hardware. The difference in speed in the case of Icon is relatively minor.

An interpreter offers a number of advantages over the generation of machine code that offset the small loss of efficiency. The main advantage is that the interpreter gets into execution very quickly, since it does not require a loading phase to resolve assembly-language references to library routines. Icode files also are much smaller than the executable binary files produced by a loader, since the run-time library does not need to be included in them. Instead, only one sharable copy of the run-time system needs to be resident in memory when Icon is executing.

## 3.3 THE TRANSLATOR

The translator that produces ucode is relatively conventional. It is written entirely in C and is independent of the architecture of the target machine on which Icon runs. Ucode is portable from one target machine to another.

The translator consists of a lexical analyzer, a parser, a code generator, and a few support routines. The lexical analyzer converts a source-language program into a stream of tokens that are provided to the parser as they are needed. The parser generates abstract syntax trees on a per-procedure basis. These abstract syntax trees are in turn processed by the code generator to produce ucode. The parser is generated automatically by Yacc from a grammatical specification. Since the translator is relatively conventional and the techniques that it uses are described in detail elsewhere (Aho, Sethi, and Ullman 1985), it is not discussed here.

There is one aspect of lexical analysis that deserves mention. The body of an Icon procedure consists of a series of expressions that are separated by semicolons. However, these semicolons usually do not need to be provided explicitly, as illustrated by examples in Chapter 2. Instead, the lexical analyzer performs semicolon insertion. If a line of a program ends with a token that is legal for ending an expression, and if the next line begins with a token that is legal for beginning an expression, the lexical analyzer generates a semicolon token between the lines. For example, the two lines

```
i := j + 3
write(i)
```

are equivalent to

```
i := j + 3;
write(i)
```

since an integer literal is legal at the end of an expression and an identifier is legal at the beginning of an expression.

If an expression spans two lines, the place to divide it is at a token that is not legal at the end of a line. For example,

```
s1 := s2 ||
s3
```

is equivalent to

```
s1 := s2 || s3
```

No semicolon is inserted, since || is not legal at the end of an expression.

### 3.4 THE LINKER

The linker reads ucode files and writes icode files. An icode file consists of an executable header that loads the run-time system, descriptive information about the file, operation codes and operands, and data specific to the program. The linker, like the translator, is written entirely in C. While conversion of ucode to icode is largely a matter of reformatting, the linker performs two other functions.

### 3.4.1 Scope Resolution

The scope of an undeclared identifier in a procedure depends on global declarations (explicit or implicit) in the program in which the procedure occurs. Since the translator in general operates on only one module of a program, it cannot resolve the scope of undeclared identifiers, because not all global scope information is contained in any one module. The linker, on the other hand, processes all the modules of a program, and hence it has the task of resolving the scope of undeclared identifiers.

An identifier may be global for several reasons:

- As the result of an explicit global declaration.
- As the name in a record declaration.
- As the name in a procedure declaration.
- As the name of a built-in function.

If an identifier with no local declaration falls into one of these categories, it is global. Otherwise it is local.

### 3.4.2 Construction of Run-Time Structures

A number of aspects of a source-language Icon program are represented at run time by various data structures. These structures are described in detail in subsequent chapters. They include procedure blocks, strings, and blocks for cset and real literals that appear in the program.

This data is represented in ucode in a machine-independent fashion. The linker converts this information into binary images that are dependent on the architecture of the target computer.

## 3.5 THE RUN-TIME SYSTEM

Most of the interesting aspects of the implementation of Icon reside in its run-time system. This run-time system is written mostly in C, although there are a few lines of assembly-language code for checking for arithmetic overflow and for co-expressions. The C portion is mostly machine-independent and portable, although some machine-specific code is needed for some idiosyncratic computer architectures and to interface some operating-system environments.

There are two main reasons for concentrating the implementation in the run-time system:

- Some features of Icon do not lend themselves to translation directly into executable code for the target machine, since there is no direct image for them in the target-machine architecture. The target machine code necessary to carry out these operations therefore is too large to place in line; instead, it is placed in library routines that are called from in-line code. Such features range from operations on structures to string scanning.
- Operations that cannot be determined at translation time must be done at run time. Such operations range from type checking to storage allocation and garbage collection.

The run-time system is logically divided into four main parts: initialization and termination routines, the interpreter, library routines called by the interpreter, and support routines called by library routines.

**Initialization and Termination Routines.** The initialization routine sets up regions in which objects created at run time are allocated. It also initializes some structures that are used during program execution. Once these tasks are completed, control is transferred to the Icon interpreter.

When a program terminates, either normally or because of an error, termination routines flush output buffers and return control to the operating system.

**The Interpreter.** The interpreter analyzes icode instructions and their operands and performs corresponding operations. The interpreter is relatively simple, since most complex operations are performed by library routines. The interpreter itself is described in Chapter 8.

**Library Routines.** Library routines are divided into three categories, depending on the way they are called by the interpreter: routines for Icon operators, routines for Icon built-in functions, and routines for complicated virtual machine instructions.

The meanings of operators are known to the translator and linker, and hence they can be called directly. On the other hand, the meanings of functions cannot be determined until they are executed, and hence they are called indirectly.

**Support Routines.** Support routines include storage allocation and garbage collection, as well as type checking and conversion. Such routines typically are called by library routines, although some are called by other support routines.

RETROSPECTIVE: Superficially, the implementation of Icon appears to be conventional. An Icon program is translated and linked to produce an executable binary file. The translator and linker *are* conventional, except that they generate code and data structures for a virtual machine instead of for a specific computer.

The run-time system dominates the implementation and plays a much larger role than is played by run-time systems in conventional implementations. This run-time system is the focus of the remainder of this book.

## EXERCISES

**3.1** Explain why there is only a comparatively small difference in execution times between a version of Icon that generates assembly-language code and one that generates virtual machine code that is interpreted.

**3.2** List all the tokens in the Icon grammar that are legal as the beginning of an expression and as the end of an expression. Are there any tokens that are legal as both? As neither?

**3.3** Is a semicolon inserted by the lexical analyzer between the following two program lines?

```
s1 := s2
|| s3
```

**3.4** Is it possible for semicolon insertion to introduce syntactic errors into a program that would be syntactically correct without semicolon insertion?

**3.5** What would be the advantages and disadvantages of merging the Icon translator and linker into a single program?

CHAPTER 4

# Values and Variables

PERSPECTIVE: No feature of the Icon programming language has a greater impact on the implementation than untyped variables—variables that have no specific type associated with them. This feature originated in Icon's predecessors as a result of a desire for simplicity and flexibility.

The absence of type declarations reduces the amount that a programmer has to learn and remember. It also makes programs shorter and (perhaps) easier to write. The flexibility comes mainly from the support for heterogeneous aggregates. A list, for example, can contain a mixture of strings, integers, records, and other lists. There are numerous examples of Icon programs in which this flexibility leads to programming styles that are concise and simple. Similarly, ''generic'' procedures, whose arguments can be of any type, often are useful, especially for modeling experimental language features.

While these facilities can be provided in other ways, such as by C's union construct, Icon provides them by the *absence* of features, which fits with the philosophy of making it easy to write good programs rather than hard to write bad ones.

The other side of the coin is that the lack of type declarations for variables makes it impossible for the translator to detect most type errors and defers type checking until the program is executed. Thus, a check that can be done only once at translation time in a language with a strong compile-time type system must be done repeatedly during program execution in Icon. Furthermore, just as the Icon translator cannot detect most type errors, a person who is writing or reading an Icon program does not have type declarations to help clarify the intent of the program.

Icon also converts arguments to the expected type where possible. This feature is, nevertheless, separable from type checking; Icon could have the latter without the former. However, type checking and conversion are naturally intertwined in the implementation.

As far as the implementation is concerned, untyped variables simplify the translator and complicate the run-time system. There is little the translator can do about types. Many operations are polymorphic, taking arguments of different types and sometimes performing significantly different computations, depending on those types. Many types are convertible to others. Since procedures are data values and may change meaning during program execution, there is nothing the translator can know about them. For this reason, the translator does not attempt any type checking or generate any code for type checking or conversion. All such code resides in the run-time routines for the functions and operations themselves.

There is a more subtle way in which untyped variables influence the implementation. Since any variable can have any type of value at any time, and can have different types of values at different times, all values must be the same size. Furthermore, Icon's rich repertoire of data types includes values of arbitrary size—lists, tables, procedures, and so on.

The solution to this problem is the concept of a *descriptor*, which either contains the data for the value, if it is small enough, or else contains a pointer to the data if it is too large to fit into a descriptor. The trick, then, is to design descriptors for all of Icon's data types, balancing considerations of size, ease of type testing, and efficiency of accessing the actual data.

## 4.1 DESCRIPTORS

Since every Icon value is represented by a descriptor, it is important that descriptors be as small as possible. On the other hand, a descriptor must contain enough information to determine the type of the value that it represents and to locate the actual data. Although values of some types cannot possibly fit into any fixed-size space, it is desirable for frequently used, fixed-sized values, such as integers, to be stored in their descriptors. This allows values of these types to be accessed directly and avoids the need to provide storage elsewhere for such values.

If Icon were designed to run on only one kind of computer, the size and layout of the descriptor could be tailored to the architecture of the computer. Since the implementation is designed to run on a wide range of computer architectures, Icon takes an approach similar to that of C. Its descriptor is composed of "words," which are closely related to the concept of a word on the computer on which Icon is implemented. One word is not large enough for a descriptor that must contain both type information and an integer or a pointer. Therefore, a descriptor consists of two words, which are designated as the *d-word* and the *v-word*, indicating that the former contains descriptive information, while the latter contains the value

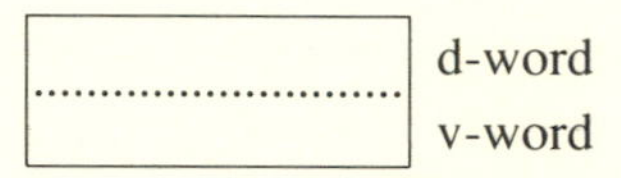


The dotted line between the two words of a descriptor is provided for readability. A descriptor is merely two words, and the fact that these two words constitute a descriptor is a matter of context.

The v-word of a descriptor may contain either a value, such as an integer, or a pointer to other data. In C terms, the v-word may contain a variety of types, including both ints and pointers. On many computers, C ints and C pointers are the same size. For some computers, however, C compilers have a large-memory-model option in which integers are 16 bits long, allowing efficient arithmetic, while pointers are 32 bits long, allowing access to a large amount of memory. In this situation, C longs are the same size as C pointers. There are other models, as well as computers with other word sizes, but the main considerations in the implementation of Icon are the accommodation of computers with 16- and 32-bit words and the large-memory model, in which pointers are larger than integers. In the large-memory model, a v-word must accommodate the largest of the types.

The d-words of descriptors contain a type code (a small integer) in their least significant bits and flags in their most significant bits. There are twelve type codes that correspond to source-language data types:

| *data type* | *type code* |
|---|---|
| null | null |
| integer | integer or long |
| real number | real |
| cset | cset |
| file | file |
| procedure | proc |
| list | list |
| set | set |
| table | table |
| record | record |
| co-expression | coexpr |

Other type codes exist for internal objects, which are on a par with source-language objects, from an implementation viewpoint, but which are not visible at the source-language level. The actual values of these codes are not important, and they are indicated in diagrams by their type code names.

### 4.1.1 Strings

There is no type code for strings. They have a special representation in which the d-word contains the length of the string (the number of characters in it) and the v-word points to the first character in the string:

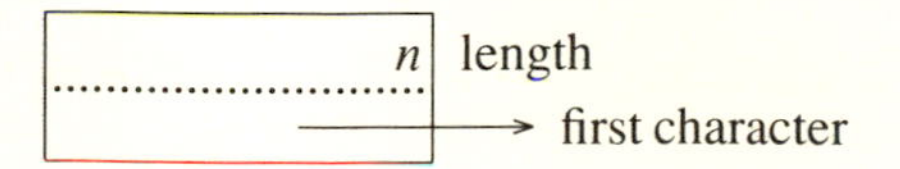


String descriptors are called *qualifiers*. In order to make qualifiers more intelligible in the diagrams that follow, a pointer to a string is followed by the string in quotation marks rather than by an address. For example, the qualifier for "hello" is depicted as

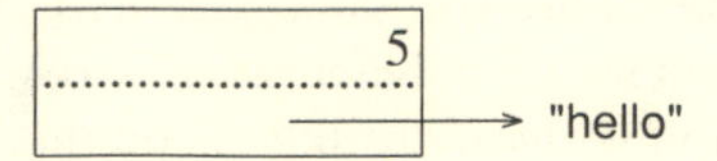


In order to distinguish qualifiers from other descriptors with type codes that might be the same as a string length, all descriptors that are not qualifiers have an n flag in the most significant bit of the d-word. The d-words of qualifiers do not have this n flag, and string lengths are restricted to prevent their overflow into this flag position in situations where words are only 16 bits long.

### 4.1.2 The Null Value

A descriptor for the null value has the form

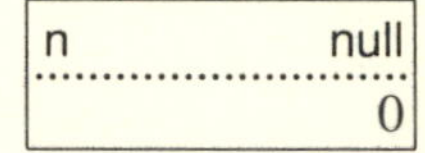


As explained previously, the n flag occurs in this and all other descriptors that are not qualifiers so that strings can be easily and unambiguously distinguished from all other kinds of values. The value in the v-word could be any constant value, but zero is useful and easily identified—and suggests "null."

### 4.1.3 Integers

Icon supports 32-bit integers, regardless of the computer on which it is implemented. Such integers therefore are either C ints or longs, depending on the computer architecture. On computers with 32-bit ints, the value of an Icon integer is stored in the v-word of its descriptor. For example, the integer 13570 is represented by

| n integer |
|---:|
| 13570 |

Note that the n flag distinguishes this descriptor from a string whose first character might be at the address 13570 and whose length might have the same value as the type code for integer.

On computers with 16-bit ints, an Icon integer that fits in 16 bits also is stored in the v-word of a descriptor. An integer that is too large to fit into a word is stored in a block that is pointed to by the v-word, as illustrated in the next section. The two representations of integers are distinguished by different internal type codes: integer for integers that are contained in the v-words of their descriptors and long for integers that are contained in blocks pointed to by the v-words of their descriptors. Thus, there are two internal types for one source-language data type.

## 4.2 BLOCKS

All other types of Icon data are represented by descriptors with v-words that point to blocks of words. These blocks have a comparatively uniform structure that is designed to facilitate their processing during garbage collection.

The first word of every block, called its *title*, contains a type code. This type code is the same code that is in the type-code portion of the d-word of a descriptor that points to the block. Some blocks are fixed in size for all values of a given type. For example, on a computer with 16-bit words, the source-language integer 80,000 is stored in a large integer block:

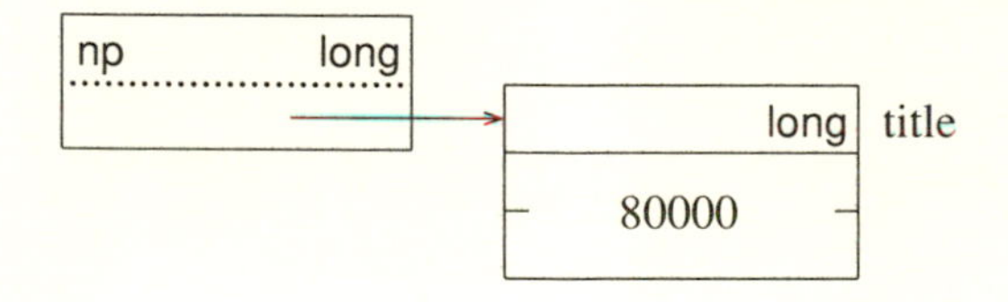


The p flag in the descriptor indicates that the v-word contains a pointer to a block.

Blocks of some other types, such as record blocks, vary in size from value to value, but any one block is fixed in size and never grows or shrinks. If the type code in the title does not determine the size of the block, the second word in the block contains its size in bytes. In the diagrams that follow, the sizes of blocks are given for computers with 32-bit words. The diagrams would be slightly different for computers with 16-bit words.

Records, which differ in size depending on how many fields they have, are examples of blocks that contain their sizes. For example, given the record declaration

```
record complex(r, i)
```

and

point := complex(1,3)

the value of point is

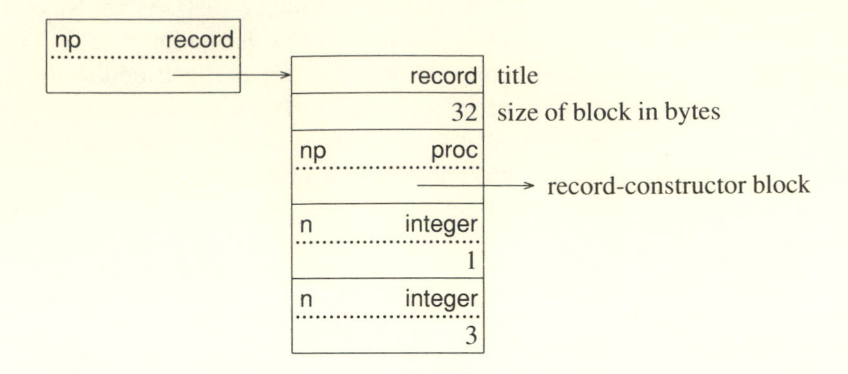


The record-constructor block contains information that is needed to resolve field references.

On the other hand, with the declaration

record term(value, code, count)

and

word := term("chair", "noun", 4)

the value of word is:

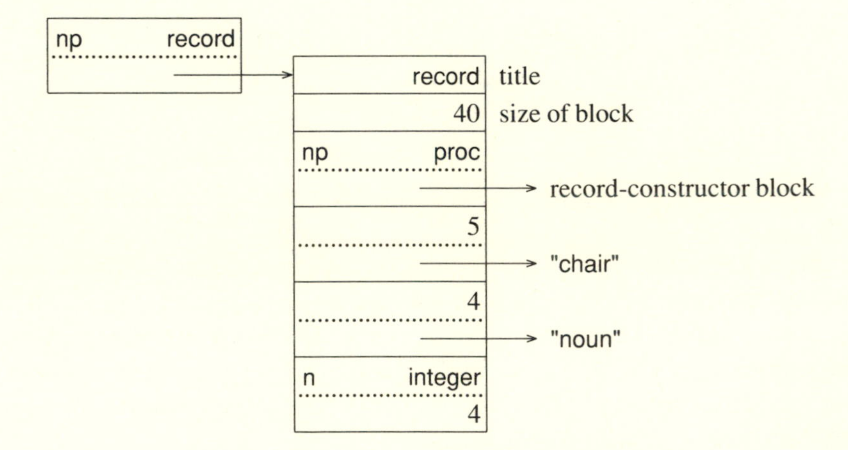

As illustrated by these examples, blocks may contain descriptors as well as non-descriptor data. Non-descriptor data comes first in the block, followed by any descriptors, as illustrated by the preceding figure. The location of the first descriptor in a block is constant for all blocks of a given type, which facilitates garbage collection.

Blocks for the remaining types are described in subsequent chapters.

## 4.3 VARIABLES

Variables are represented by descriptors, just as values are. This representation allows values and variables to be treated uniformly in terms of storage and access. Variables for identifiers point to descriptors for the corresponding values. Variables always point to descriptors for values, never to other variables. For example, if

```
s := "hello"
```

then a variable for s has the form

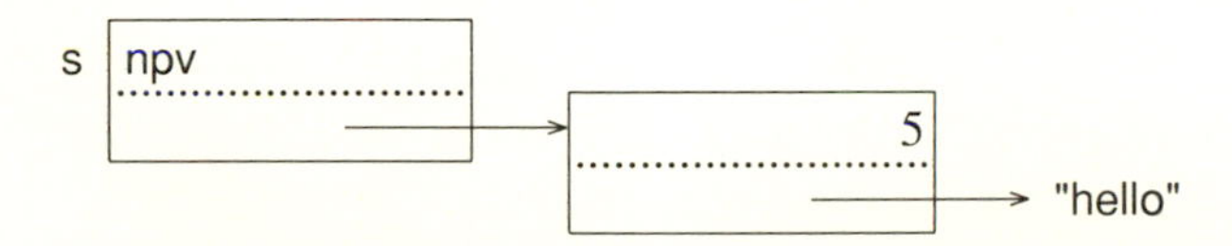


The v flag distinguishes descriptors for variables from descriptors for values.

The values of local identifiers are kept on a stack, while the values of global and static identifiers are located at fixed places in memory. Variables that point to the values of identifiers are created by icode instructions that correspond to the use of the identifiers in the program.

Some variables, such as record field references, are computed. A variable that references a value in a data structure points directly to the descriptor for the value. The least-significant bits of the d-word for such a variable contain the offset, in *words*, of the value descriptor from the top of the block in which the value is contained. This offset is used by the garbage collector. The use of words, rather than bytes, allows larger offsets, which is important for computers with 16-bit words. For example, the variable word.count for the record given in the preceding section is

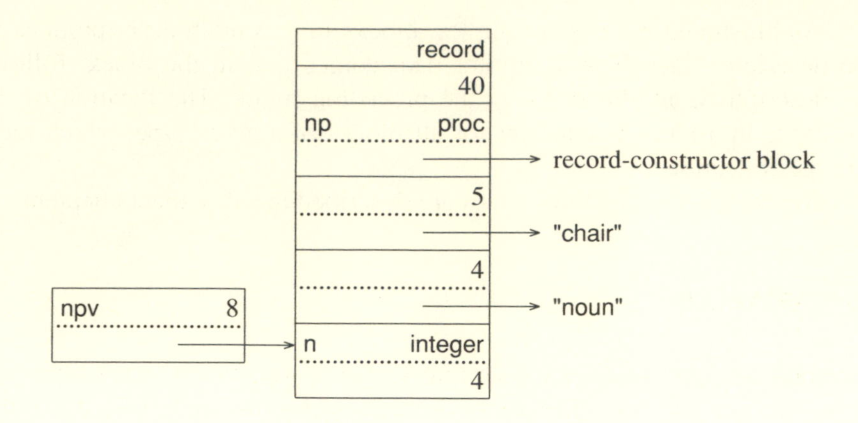


The variable points directly to the value rather than to the title of the block so that access to the value is more efficient. Note that the variable word.count cannot be determined at translation time, since the type of word is not known then and different record types could have count fields in different positions.

### 4.3.1 Operations on Variables

There are two fundamentally different contexts in which a variable can be used: *dereferencing* and *assignment*.

Suppose, as shown previously, that the value of the identifier s is the string "hello". Then a variable descriptor that points to the value of s and the corresponding value descriptor for "hello" have the following relationship:

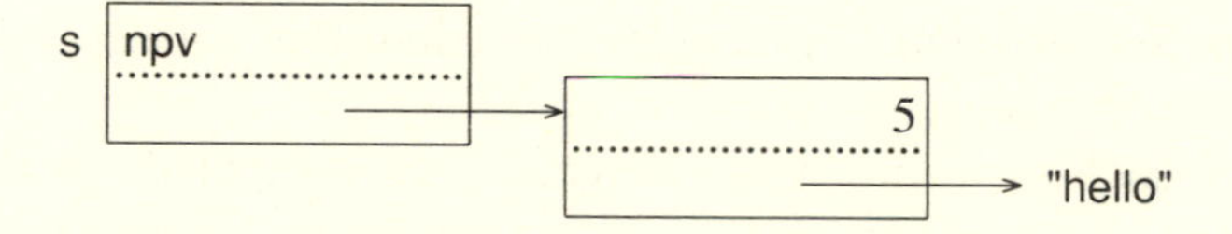


In an expression such as write(s), s is dereferenced by fetching the descriptor pointed to by the v-word of the variable.

In the case of assignment, as in

```
s := 13570
```

the value descriptor pointed to by the v-word of the variable descriptor is changed:

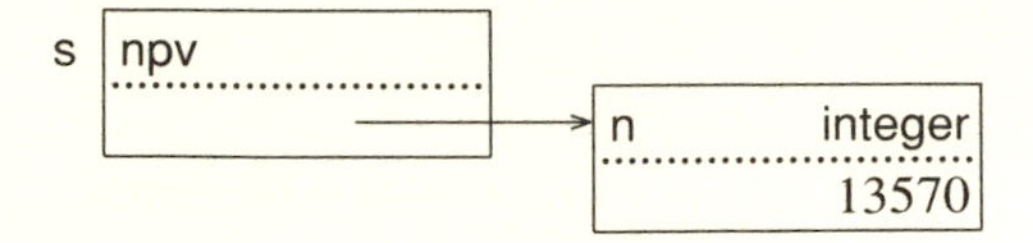


These operations on variables correspond to indirect load and store instructions of a typical computer.

### 4.3.2 Trapped Variables

Icon has several variables with special properties that complicate assignment and dereferencing. Consider, for example, the keyword &trace. Its value must always be an integer. Consequently, in an assignment such as

&trace := *expr*

the value produced by *expr* must be checked to be sure that it is an integer. If it is not, an attempt is made to convert it to an integer, so that in

&trace := "1"

the value assigned to &trace is the integer 1, not the string "1".

There are four keyword variables that require special processing for assignment: &trace, &random, &subject, and &pos. The keyword &random is treated in essentially the same way that &trace is. Assignment to &subject requires a string value and has the side effect of assigning the value 1 to &pos. Assignment to &pos is even more complicated: not only must the value assigned be an integer, but if it is not positive, it must also be converted to the positive equivalent with respect to the length of &subject. In any event, if the value in the assignment to &pos is not in the range of &subject, the assignment fails. Dereferencing these keywords, on the other hand, requires no special processing.

A naive way to handle assignment to these keywords is to check every variable during assignment to see whether it is one of the four that requires special processing. This would place a significant computational burden on every assignment. Instead, Icon divides variables into two classes: *ordinary* and *trapped*. Ordinary variables point to their values as illustrated previously and require no special processing. Trapped variables, so called because their processing is "trapped," are distinguished from ordinary variables by a t flag. Thus, assignment only has to check a single flag to separate the majority of variables from those that require special processing.

A trapped-variable descriptor for a keyword points to a block that contains the value of the keyword, its string name, and a pointer to a C function that is called when assignment to the keyword is made. For example, the trapped variable for &trace is:

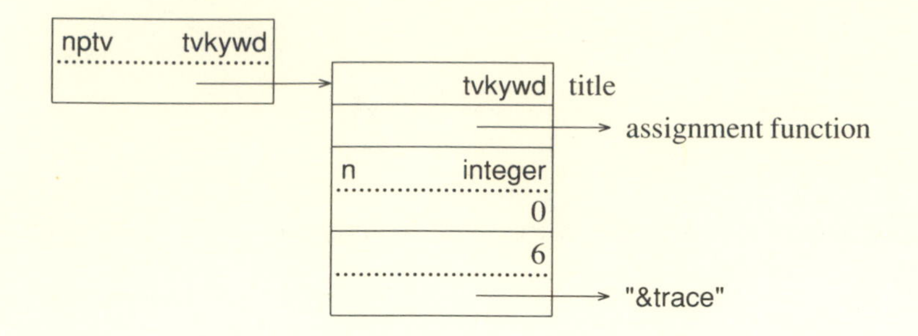


It is worth noting that the more conventional approach to handling the problem of assignment to keywords is to compile special code if a keyword occurs in an assignment context. It is not always possible, however, to determine the context in which a variable is used in Icon. Consider a procedure of the form

```
procedure diagnose(s)
        .
        .
        .
  return &trace
end
```

The semantics of Icon dictate that the result returned in this case should be a variable, not just its value, so that it is possible to write an expression such as

```
diagnose(s) := 10
```

which has the effect of assigning the value 10 to &trace.

The translator has no way of knowing that an assignment to the call diagnose(s) is equivalent to an assignment to &trace. In fact, the translator cannot even determine that the value of diagnose will be a function when the previous assignment is performed, much less that it will be the procedure given earlier.

Thus, the trapped-variable mechanism provides a way to handle, uniformly, all the situations in which such a keyword can be used.

## 4.4 DESCRIPTORS AND BLOCKS IN C

Descriptors and blocks of data are described and depicted abstractly in the previous sections of this chapter. In order to understand the implementation of some aspects of Icon, it is helpful to examine the C code that actually defines and manipulates data.

The following sections illustrate typical C declarations for the structures used in the implementation of Icon. Some of the terminology and operations that appear frequently in the C code are included as well. Other operations are introduced in subsequent chapters, as they are needed.

### 4.4.1 Descriptors

As mentioned in Sec. 4.1, for C compilers in which ints and pointers are the same size, the size of a word is the size of an int, while if pointers are larger than ints, the size of a word is the size of a long. The difference between these two models of memory is handled by typedefs under the control of conditional compilation. Two constants that characterize the sizes are defined: IntSize and PtrSize. If these sizes are different, the constant MixedSizes is defined:

```
#if IntSize != PtrSize
#define MixedSizes
#endif
```

This constant is used to select appropriate definitions for signed and unsigned words:

```
#ifdef MixedSizes
typedef long word;
typedef unsigned long uword;
#else
typedef int word;
typedef unsigned int uword;
#endif
```

A descriptor is declared as a structure:

```
struct descrip {                /* descriptor */
   word dword;                  /*    type field */
   union {
      word integr;              /*    integer value */
      char *sptr;               /*    pointer to character string */
      union block *bptr;        /*    pointer to a block */
      struct descrip *dptr;     /*    pointer to a descriptor */
      } vword;
   };
```

The v-word of a descriptor is a union that reflects its various uses: an integer, a pointer to a string, a pointer to a block, or a pointer to another descriptor (in the case of a variable).

### 4.4.2 Blocks

Each block type has a structure declaration. For example, the declaration for record blocks is

```
struct b_record {                 /* record block */
   word title;                    /*    T_Record */
   word blksize;                  /*    size of block */
   struct descrip recdesc;        /*    descriptor for record constructor */
   struct descrip fields[1];      /*    fields */
   };
```

Blocks for records vary in size, depending on the number of fields declared for the record type. The size of 1 in

```
struct descrip fields[1];
```

is provided to satisfy the C compiler. Actual blocks for records are constructed at run time in a region that is managed by Icon's storage allocator. Such blocks conform to the previous declaration, but the number of fields varies. The declaration provides a means of accessing portions of such blocks from C.

The declaration for keyword trapped-variable blocks is

```
struct b_tvkywd {                 /* keyword trapped variable block */
   word title;                    /*    T_Tvkywd */
   int (*putval) ();              /*    assignment function for keyword */
   struct descrip kyval;          /*    keyword value */
   struct descrip kyname;         /*    keyword name */
   };
```

Note that the title fields of b_record and b_tvkywd contain type codes, as indicated in previous diagrams. The second field of b_record is a size as mentioned previously, but b_tvkywd has no size field, since all keyword trapped-variable blocks are the same size, which therefore can be determined from their type.

The block union given in the declaration of descrip consists of a union of all block types:

```
union block {                    /* general block */
   struct b_int longint;
   struct b_real realblk;
   struct b_cset cset;
   struct b_file file;
   struct b_proc proc;
   struct b_list list;
   struct b_lelem lelem;
   struct b_table table;
   struct b_telem telem;
   struct b_set set;
   struct b_selem selem;
   struct b_record record;
   struct b_tvkywd tvkywd;
   struct b_tvsubs tvsubs;
   struct b_tvtbl tvtbl;
   struct b_coexpr coexpr;
   struct b_refresh refresh;
   };
```

Note that there are several kinds of blocks in addition to those that correspond to source-language data types.

### 4.4.3 Defined Constants

The type codes are defined symbolically:

```
#define T_Null       0
#define T_Integer    1
#define T_Long       2
#define T_Real       3
#define T_Cset       4
#define T_File       5
#define T_Proc       6
#define T_List       7
#define T_Table      8
#define T_Record     9
#define T_Telem     10
#define T_Lelem     11
#define T_Tvsubs    12
#define T_Tvkywd    13
#define T_Tvtbl     14
```

```
#define T_Set        15
#define T_Selem      16
#define T_Refresh    17
#define T_Coexpr     18
```

The type codes in diagrams are abbreviated, as indicated by previous examples.

The defined constants for d-word flags are

```
n    F_Nqual
p    F_Ptr
v    F_Var
t    F_Tvar
```

The values of these flags depend on the word size of the computer.

The d-words of descriptors are defined in terms of flags and type codes:

```
#define D_Null       (T_Null | F_Nqual)
#define D_Integer    (T_Integer | F_Nqual)
#define D_Long       (T_Long | F_Ptr | F_Nqual)
#define D_Real       (T_Real | F_Ptr | F_Nqual)
#define D_Cset       (T_Cset | F_Ptr | F_Nqual)
#define D_File       (T_File | F_Ptr | F_Nqual)
#define D_Proc       (T_Proc | F_Ptr | F_Nqual)
#define D_List       (T_List | F_Ptr | F_Nqual)
#define D_Table      (T_Table | F_Ptr | F_Nqual)
#define D_Set        (T_Set | F_Ptr | F_Nqual)
#define D_Selem      (T_Selem | F_Ptr | F_Nqual)
#define D_Record     (T_Record | F_Ptr | F_Nqual)
#define D_Telem      (T_Telem | F_Ptr | F_Nqual)
#define D_Lelem      (T_Lelem | F_Ptr | F_Nqual)
#define D_Tvsubs     (T_Tvsubs | D_Tvar)
#define D_Tvtbl      (T_Tvtbl | D_Tvar)
#define D_Tvkywd     (T_Tvkywd | D_Tvar)
#define D_Coexpr     (T_Coexpr | F_Ptr | F_Nqual)
#define D_Refresh    (T_Refresh | F_Ptr | F_Nqual)

#define D_Var        (F_Var | F_Nqual | F_Ptr)
#define D_Tvar       (D_Var | F_Tvar)
```

As indicated previously, flags, type codes, and d-words are distinguished by the prefixes F_, T_, and D_, respectively.

### 4.4.4 C Coding Conventions

A number of conventions are used in the C routines for the run-time system to reduce detail and to focus on the way that Icon data is organized. Some of these are illustrated by the C function for the Icon operator *x, which produces the size of x:

```
OpDcl(size, 1, "*")
   {
   char sbuf[MaxCvtLen];

   Arg0.dword = D_Integer;
   if (Qual(Arg1)) {
      /*
       * If Arg1 is a string, return the length of the string.
       */
      IntVal(Arg0) = StrLen(Arg1);
      }

   else {
      /*
       * Arg1 is not a string.  For most types, the size is in the size
       *  field of the block.
       *  structure.
       */
      switch (Type(Arg1)) {
         case T_List:
            IntVal(Arg0) = BlkLoc(Arg1)->list.size;
            break;

         case T_Table:
            IntVal(Arg0) = BlkLoc(Arg1)->table.size;
            break;

         case T_Set:
            IntVal(Arg0) = BlkLoc(Arg1)->set.size;
            break;

         case T_Cset:
            IntVal(Arg0) = BlkLoc(Arg1)->cset.size;
            break;
                        .
                        .
                        .
```

```
            default:
               /*
                * Try to convert it to a string.
                */
               if (cvstr(&Arg1, sbuf) == CvtFail)
                  runerr(112, &Arg1);           /* no notion of size */
               IntVal(Arg0) = StrLen(Arg1);
            }
      }
   Return;
   }
```

OpDcl is a macro that performs several operations. One of these operations is to provide a C function declaration. Since the function is called by the interpreter, the header is somewhat different from what it would be if size were called directly. The details are described in Chapter 8.

By convention, the arguments of the Icon operation are referred to via Arg1, Arg2, ... . The result that is produced for an operator is left in Arg0 rather than being given as an argument of return. Thus, in the case of *x, the value of x is in Arg1 and the returned size is placed in Arg0.

First, the d-word of Arg0 is set to D_Integer, since the returned value is an integer. Next, there is a test to determine if Arg1 is a qualifier. Qual is a macro that is defined as

```
#define Qual(d)     (!((d).dword & F_Nqual))
```

If Arg1 is a qualifier, its length is placed in the v-word of Arg0, using the macros IntVal and StrLen, which are defined as

```
#define IntVal(d)    ((d).vword.integr)
#define StrLen(d)    ((d).dword)
```

If Arg1 is not a qualifier, then the size depends on the type. The macro Type isolates the type code

```
#define Type(d)     ((d).dword & TypeMask)
```

where the value of TypeMask is 63, providing considerable room for additions to Icon's 19 internal types.

For most Icon types that are represented by blocks, their source-language size is contained in their size field. The macro BlkLoc accesses a pointer in the v-field of a descriptor and is defined as

```
#define BlkLoc(d)   ((d).vword.bptr)
```

If the type is not one of these, the final task is an attempt to convert Arg1 to a string. The support routine cvstr does this, using the buffer sbuf provided by size. The value of Arg1 is changed accordingly; note that its address is provided to cvstr. A fixed-sized buffer can be used, since there is a limit to the size of a string that can be obtained by converting other types. This limit is 256, which is reached only for conversion of &cset. The conversion may fail, as for *&null, which is signalled by the return value NULL from cvstr. In this case, program execution is terminated with a run-time error message, using runerr. If the conversion is successful, the size is placed in the v-word of Arg0, as is the case if Arg1 was a qualifier originally. Note that the original test for a qualifier could be replaced by a call to cvstr, and the call to cvstr in the default of the switch statement could be eliminated. The code is written the way it is for efficiency, avoiding the call to cvstr in the common case that the argument is a string. It is worth noting that a special case is needed for strings, since a qualifier has no type code and a test for a string cannot be included in the switch statement.

The macro Return returns from the function and signals the interpreter that a result has been produced.

RETROSPECTIVE: Descriptors provide a uniform way of representing Icon values and variables. Since descriptors for all types of data are the same size, there are no problems with assigning different types of values to a variable—they all fit.

The importance of strings is reflected in the separation of descriptors into two classes—qualifiers and nonqualifiers—by the n flag. The advantages of the qualifier representation for strings are discussed in Chapter 5.

It is comparatively easy to add a new type to Icon. A new type code is needed to distinguish it from other types. If the possible values of the new type are small enough to fit into the v-word, as is the case for integers, no other data is needed. For example, the value of a character data type could be contained in its descriptor. For types that have values that are too large to fit into a v-word, pointers to blocks containing the data are placed in the v-words instead. Lists, sets, and tables are examples of data types that are represented this way. See Chapters 6 and 7.

## EXERCISES

**4.1** Give examples of Icon programs in which heterogeneous aggregates are used in significant ways.

**4.2** Design a system of type declarations for Icon so that the translator could do type checking. Give special consideration to aggregates, especially those that may change in size during program execution. Do this from two perspectives: (a) changing the semantics of Icon as little as possible, and (b) maximizing the type checking that can be done by the translator at the expense of flexibility in programming.

**4.3** Suppose that functions in Icon were not first-class values and that their meanings were bound at translation time. How much could the translator do in the way of error checking?

**4.4** Compile a list of all Icon functions and operators. Are there any that do not require argument type checking? Are there any that require type checking but not conversion? Identify those that are polymorphic. For the polymorphic ones, identify the different kinds of computations that are performed depending on the types of the arguments.

**4.5** Compose a table of all type checks and conversions that are required for Icon functions and operators.

**4.6** To what extent would the implementation of Icon be simplified if automatic type conversion were not supported? How would this affect the programmer?

**4.7** Why is it desirable for string qualifiers not to have flags and for all other kinds of descriptors to have flags indicating they are not qualifiers, rather than the other way around?

**4.8** Is the n flag that distinguishes string qualifiers from all other descriptors really necessary? If not, explain how to distinguish the different types of descriptors without this flag.

**4.9** On computers with extremely limited address space, two-word descriptors may be impractically large. Describe how one-word descriptors might be designed, discuss how various types might be represented, and describe the ramifications for storage utilization and execution speed.

**4.10** Identify the diagrams in this chapter that would be different if they were drawn for a computer with 16-bit words. Indicate the differences.

**4.11** There is nothing in the nature of keywords that requires them to be processed in a special way for assignment but not for dereferencing. Invent a new keyword that is a variable that requires processing when it is dereferenced. Show how to generalize the keyword trapped-variable mechanism to handle such cases.

**4.12** List all the syntactically distinct cases in which the translator can determine whether a keyword variable is used in an assignment or dereferencing context.

**4.13** What would be gained if special code were compiled for those cases in which the context for keyword variables could be determined?

CHAPTER 5

# Strings and Csets

PERSPECTIVE: Several aspects of strings as a language feature in Icon have a strong influence on how they are handled by the implementation. First of all, strings are the most frequently used type of data in the majority of Icon programs. The number of different strings and the total amount of string data often are large. Therefore, it is important to be able to store and access strings efficiently.

Icon has many operations on strings—nearly fifty of them. Some operations, such as determining the size of a string, are performed frequently. The efficiency of these operations is an important issue and influences, to a considerable extent, how strings are represented.

Icon strings may be very long. Although some limitation on the maximum length of a string may be acceptable as a compromise with the architecture of the computer on which Icon is implemented (and hence considerations of efficiency), this maximum must be so large as to be irrelevant for most Icon programs.

String lengths are determined dynamically during program execution, instead of being specified statically in declarations. Much of the advantage of string processing in Icon over other programming languages comes from the automatic management of storage for strings.

Any of the 256 8-bit ASCII characters can appear in an Icon string. Even the ''null'' character is allowed.

Several operations in Icon return substrings of other strings. Substrings tend to occur frequently, especially in programs that analyze (as opposed to synthesize) strings.

Strings in Icon are atomic—there are no operations in Icon that change the characters in existing strings. This aspect of Icon is not obvious; in fact, there are operations that appear to change the characters in strings. The atomic nature of string operations in Icon simplifies its implementation considerably. For example, assignment of a string value to a variable need not (and does not) copy the string.

The order in which characters appear is an essential aspect of strings. There are many situations in Icon, however, where several characters have the same status but where their order is irrelevant. For example, the concepts of vowels and punctuation marks depend on set membership but not on order. Csets are provided for such situations. Interestingly, many computations can be performed using csets that have nothing to do with the characters themselves (Griswold and Griswold 1983, pp. 181-191).

## 5.1 STRINGS

### 5.1.1 Representation of Strings

Although it may appear natural for the characters of a string to be stored in consecutive bytes, this has not always been so. On earlier computer architectures without byte addressing and character operations, some string-manipulation languages represented strings by linked lists of words, each word containing a single character. Such a representation seems bizarre for modern computer architectures and obviously consumes a very large amount of memory—an intolerable amount for a language like Icon.

The C programming language represents strings (really arrays of characters) by successive bytes in memory, using a zero (null) byte to indicate the end of a string. Consequently, the end of a string can be determined from the string itself, without any external information. On the other hand, determining the length of a string, if it is not already known, requires indexing through it, incrementing a counter until a null byte is found. Furthermore, and very important for a language like Icon, substrings (except terminal ones) cannot occur within strings, since every C string must end with a null byte.

Since any character can occur in an Icon string, it is not possible to use C's null-termination approach to mark ends of strings. Therefore, there is no way to detect the end of a string from the string itself, and there must be some external way to determine where a string ends. This consideration provides the motivation for the qualifier representation described in the last chapter. The qualifier provides information, external to the string itself, that delimits the string by the address of its first character and its length. Such a representation makes the computation of substrings fast and simple—and, of course, determining the length of a string is fast and independent of its length.

Note that C-style strings serve perfectly well as Icon-style strings; the null byte at the end of a C-style string can be ignored by Icon. This allows strings produced by C functions to be used by Icon. The converse is not true; in order for an Icon string to be used by C, a copy must be made with a null byte appended at the end.

Some strings are compiled into the run-time system and others, such as strings that appear as literals in a program, are contained in icode files that are loaded into memory when program execution begins. During program execution, Icon strings may be stored in work areas (usually referred to as ''buffers''). Most newly created strings, however, are allocated in a common string region.

As source-language operations construct new strings, their characters are appended to the end of those already in the string region. The amount of space allocated in the string region typically increases during program execution until the region is full, at which point it is compacted by garbage collection, squeezing out characters that are no longer needed. See Chapter 11 for details.

In the previous chapter, the string to which a qualifier points is depicted by an arrow followed by the string. For example, the string "the" is represented by the qualifier

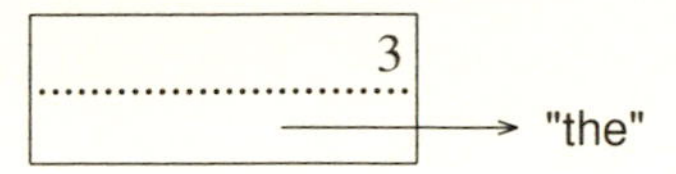


The pointer to "the" is just a notational convenience. A more accurate representation is

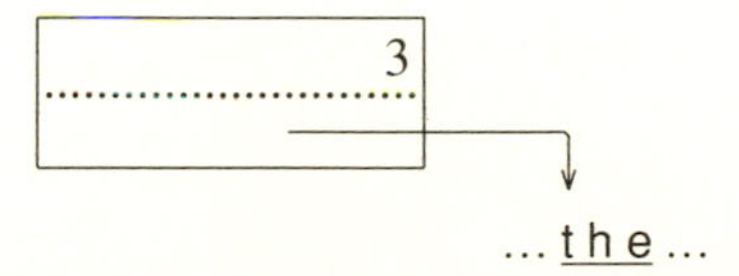


The actual value of the v-word might be 0x569a (hexadecimal), where the character t is at memory location 0x569a, the character h is at location 0x569b, and the character e is at location 0x569c.

### 5.1.2 Concatenation

In an expression such as

```
s := "hello"
```

the string "hello" is contained in data provided as part of the icode file, and a qualifier for it is assigned to s; no string is constructed. Some operations that produce strings require the allocation of new strings. Concatenation is a typical example:

```
s1 := "ab" || "cdef"
```

In this expression, the concatenation operation allocates space for six characters, copies the two strings into this space, and produces a qualifier for the result:

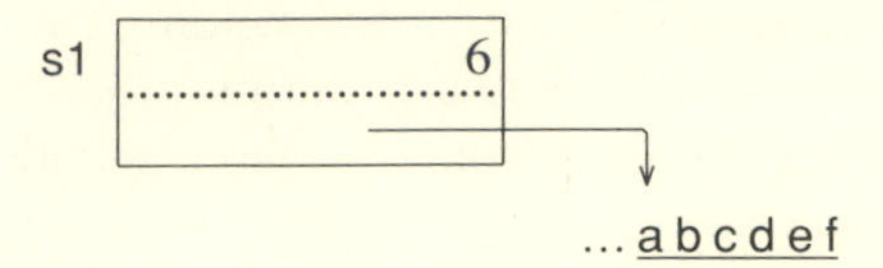


This qualifier then becomes the value of s1.

There is one important optimization in concatenation. If the first argument in a concatenation is the last string in the string region, the second argument is simply appended to the end of the string region. Thus, operations of the form

s := s || *expr*

perform less allocation than operations of the form

s := *expr* || s

Except for this optimization, no string construction operation attempts to use another instance of a string that may exist somewhere else in the string region. As a result,

```
s1 := "ab" || "c"
s2 := "a" || "bc"
```

produce two distinct strings:

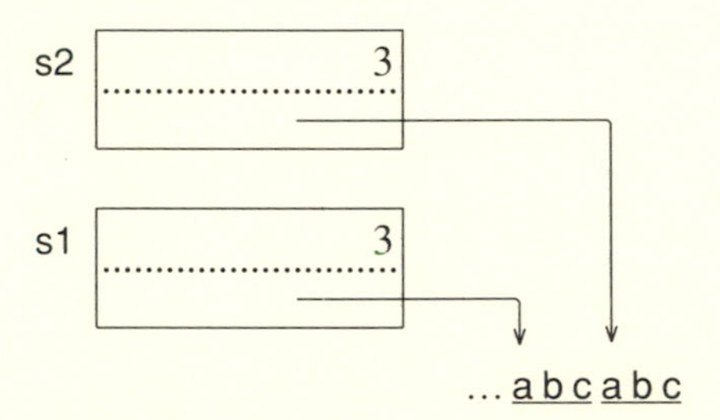


The C code for the concatenation operation is

```
OpDcl(cat, 2, "||")
   {
   char sbuf1[MaxCvtLen];        /* buffers for conversion to string */
   char sbuf2[MaxCvtLen];
   extern char *alcstr();
```

```
/*
 *  Convert arguments to strings if necessary.
 */
if (cvstr(&Arg1, sbuf1) == CvtFail)
   runerr(103, &Arg1);
if (cvstr(&Arg2, sbuf2) == CvtFail)
   runerr(103, &Arg2);

/*
 * Ensure space for the resulting string.
 */
strreq(StrLen(Arg1) + StrLen(Arg2));

if (StrLoc(Arg1) + StrLen(Arg1) == strfree)
   /*
    * The end of Arg1 is at the end of the allocated string space.
    *  Hence, Arg1 was the last string allocated. Arg1 is not
    *  copied. Instead, Arg2 is appended to the string space and
    *  the result is pointed to the start of Arg1.
    */
   StrLoc(Arg0) = StrLoc(Arg1);

else
   /*
    * Otherwise, append Arg1 to the end of the allocated
    *  string space and point the result to the start of Arg1.
    */
   StrLoc(Arg0) = alcstr(StrLoc(Arg1),StrLen(Arg1));

/*
 * Append Arg2 to the end.
 */
alcstr(StrLoc(Arg2),StrLen(Arg2));
/*
 *  Set the length of the result and return.
 */
StrLen(Arg0) = StrLen(Arg1) + StrLen(Arg2);
Return;
}
```

The function strreq(n) assures that there are at least n bytes available in the allocated string region. See Chapter 11 for details. The function alcstr(s, n) allocates n characters and copies s to that space. The global variable strfree points to the beginning of the free space at the end of the allocated string region.

### 5.1.3 Substrings

Many string operations do not require the allocation of a new string but only produce new qualifiers. For example, if the value of s1 is "abcdef", the substring formed by

```
s2 := s1[3:6]
```

does not allocate a new string but only produces a qualifier that points to a substring of s1:

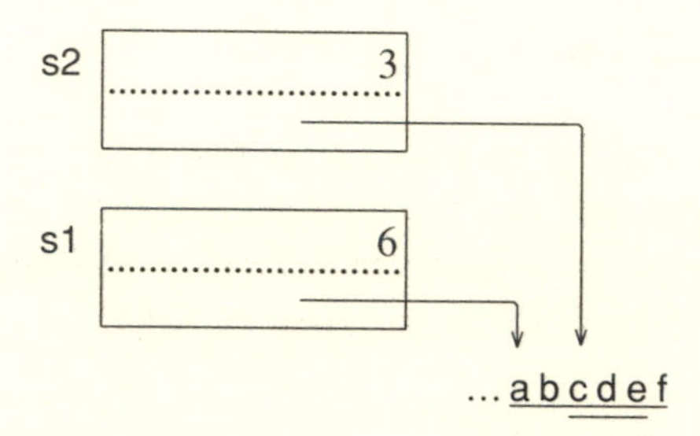


In order for Icon string values to be represented in memory by substrings, it is essential that there be no Icon operation that changes the characters inside a string. As mentioned earlier, this is the case, although it is not obvious from a cursory examination of the language. C, on the other hand, allows the characters in a string to be changed. The difference is that C considers a string to be an array of characters and allows assignment to the elements of the array, while Icon considers a string to be an indivisible atomic object. It makes no more sense in Icon to try to change a character in a string than it does to try to change a digit in an integer. Thus, if

```
i := j
```

and

```
j := j + 1
```

the value of i does not change as a result of the subsequent assignment to j. So it is with strings in Icon.

Admittedly, there are operations in Icon that *appear* to change the characters in a string. For example,

```
s1[3] := "x"
```

gives the appearance of changing the third character in s1 to "x". However, this expression is simply shorthand for

```
s1 := s1[1:2] || "x" || s1[4:0]
```

A new string is created by concatenation and a new qualifier for it is assigned to s1, as shown by

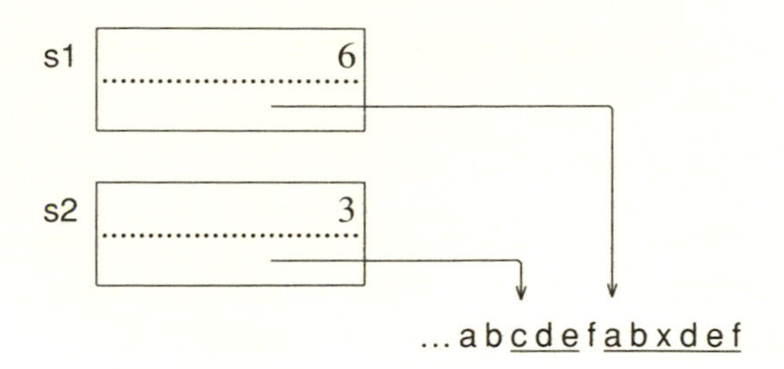


Of course, the length of the string may be increased or decreased by assignment to a substring, as in

```
s1[3] := "xxx"
s1[2:5] := ""
```

### 5.1.4 Assignment to Subscripted Strings

Expressions such as x[i] and x[i:j] represent a particular challenge in the implementation of Icon. In the first place, the translator cannot determine the type of x. In the case of x[i], there are four basic types that x may legitimately have: string, list, table, and record. Of course, any type that can be converted to a string is legitimate also. Unfortunately, the *nature* of the operation, not just the details of its implementation, depends on the type. For strings,

```
s1[3] := s2
```

replaces the third character of s1 by s2 and is equivalent to concatenation, as described previously. For lists, tables, and records,

```
x[3] := y
```

changes the third *element* of x to y—quite a different matter (see Exercise 5.5).

This problem is pervasive in Icon and only needs to be noted in passing here. The more serious problem is that even if the subscripted variable is a string, the subscripting expression has different meanings, depending on the context in which it appears.

If s is a variable, then s[i] and s[i:j] also are variables. In a dereferencing context, such as

```
write(s[2:5])
```

the result produced by s[2:5] is simply a substring of s, and the subscripting expression produces the appropriate qualifier.

Assignment to a subscripted string, as in

```
s[2:5] := "xxx"
```

is not at all what it appears to be superficially. Instead, as already noted, it is shorthand for an assignment to s:

```
s := s[1] || "xxx" || s[6:0]
```

If the translator could determine whether a subscripting expression is used in a dereferencing or assignment context, it could produce different code for the two cases. As mentioned in Sec. 2.2, however, the translator cannot always make this determination. Consequently, trapped variables are used for subscripted strings much in the way they are used for keywords. For example, if the value of s is "abcdef", the result of evaluating the subscripting expression s[2:5] is a *substring trapped variable* that has the form

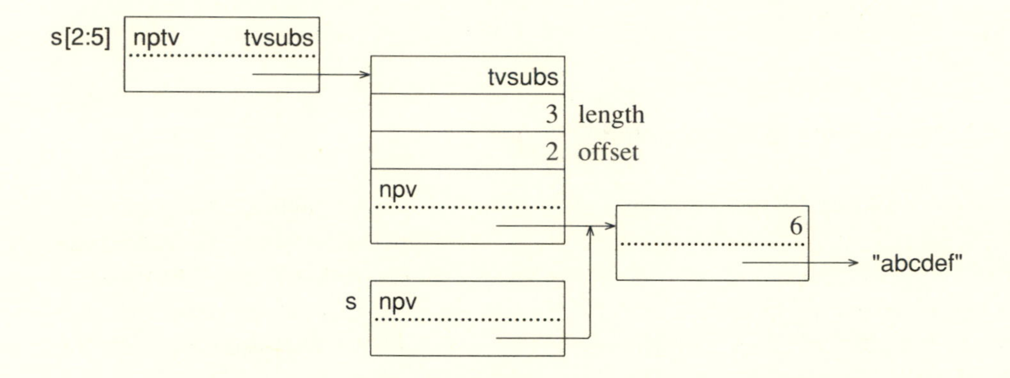


Note that both the variable for s and the variable in the substring trapped-variable block point to the same value. This makes it possible for assignment to the substring trapped variable to change the value of s.

The length and offset of the substring provide the necessary information either to produce a qualifier for the substring, in case the subscripting expression is dereferenced, or to construct a new string in case an assignment is made to the subscripting expression. For example, after an assignment such as

```
s[2:5] := "x"
```

the situation is

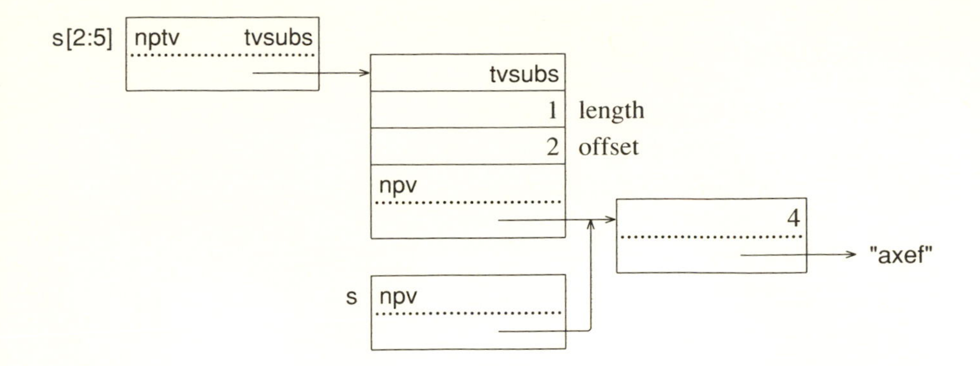


Note that the value of s has changed. The length of the subscripted portion of the string has been changed to correspond to the length of the string assigned to it. This reflects the fact that subscripting identifies the portions of the string before and after the subscripted portion ("a" and "ef", in this case). In the case of a multiple assignment to a subscripted string, only the original subscripted portion is changed. Thus, in

```
(s[2:5] := "x") := "yyyyy"
```

the final value of s is "ayyyyyef".

### 5.1.5 Mapping

String mapping is interesting in its own right, and the C function that implements it illustrates several aspects of string processing:

```
FncDcl(map, 3)
   {
   register int i;
   register word slen;
   register char *s1, *s2, *s3;
   char sbuf1[MaxCvtLen], sbuf2[MaxCvtLen], sbuf3[MaxCvtLen];
   static char maptab[256];
   extern char *alcstr();
```

```
/*
 * Arg1 must be a string; Arg2 and Arg3 default to &ucase and &lcase,
 *  respectively.
 */
if (cvstr(&Arg1, sbuf1) == CvtFail)
   runerr(103, &Arg1);
if (ChkNull(Arg2))
   Arg2 = ucase;
if (ChkNull(Arg3))
   Arg3 = lcase;

/*
 * If Arg2 and Arg3 are the same as for the last call of map,
 *  the current values in maptab can be used. Otherwise, the
 *  mapping information must information must be recomputed.
 */
if (!EqlDesc(maps2, Arg2) || !EqlDesc(maps3, Arg3)) {
   maps2 = Arg2;
   maps3 = Arg3;

   /*
    * Convert Arg2 and Arg3 to strings.  They must be of the
    *  same length.
    */
   if (cvstr(&Arg2, sbuf2) == CvtFail)
      runerr(103, &Arg2);
   if (cvstr(&Arg3, sbuf3) == CvtFail)
      runerr(103, &Arg3);
   if (StrLen(Arg2) != StrLen(Arg3))
      runerr(208, NULL);
```

```
    /*
     * The array maptab is used to perform the mapping.  First,
     *  maptab[i] is initialized with i for i from 0 to 255.
     *  Then, for each character in Arg2, the position in maptab
     *  corresponding to the value of the character is assigned
     *  the value of the character in Arg3 that is in the same
     *  position as the character from Arg2.
     */
    s2 = StrLoc(Arg2);
    s3 = StrLoc(Arg3);
    for (i = 0; i <= 255; i++)
       maptab[i] = i;
    for (slen = 0; slen < StrLen(Arg2); slen++)
       maptab[s2[slen]&0377] = s3[slen];
    }

 if (StrLen(Arg1) == 0) {
    Arg0 = emptystr;
    Return;
    }

 /*
  * The result is a string the size of Arg1; ensure that much space.
  */
 slen = StrLen(Arg1);
 strreq(slen);
 s1 = StrLoc(Arg1);

 /*
  * Create the result string, but specify no value for it.
  */
 StrLen(Arg0) = slen;
 StrLoc(Arg0) = alcstr(NULL, slen);
 s2 = StrLoc(Arg0);

 /*
  * Run through the string, using values in maptab to do the
  *  mapping.
  */
 while (slen-- > 0)
    *s2++ = maptab[(*s1++)&0377];
 Return;
 }
```

The mapping is done using the character array maptab. This array is set up by first assigning every possible character to its own position in maptab and then replacing the characters at positions corresponding to characters in s2 by the corresponding characters in s3. Note that if a character occurs more than once in s2, its last (rightmost) correspondence with a character in s3 applies.

To avoid rebuilding maptab unnecessarily, this step is bypassed if map is called with the same values of s2 and s3 as in the previous call. The global variables maps2 and maps3 are used to hold these "cached" values. The macro EqlDesc(d1, d2) tests the equivalence of the descriptors d1 and d2.

The function map is an example of a function that defaults null-valued arguments. Omitted arguments are supplied as null values. The defaults for s2 and s3 are &ucase and &lcase, respectively. Consequently,

```
map(s)
```

is equivalent to

```
map(s, &ucase, &lcase)
```

The macro ChkNull(d) tests whether or not d is null. The values of &ucase and &lcase are in the global variables ucase and lcase.

## 5.2 CSETS

Since Icon uses 8-bit characters, regardless of the computer on which it is is implemented, there are 256 different characters that can occur in csets. A cset block consists of the usual title containing the cset type code followed by a word that contains the number of characters in the cset. Next, there are words containing a total of 256 bits. Each bit represents one character, with a bit value of 1 indicating that the character is present in the cset and a bit value of 0 indicating it is absent. An example is the value of the keyword &ascii:

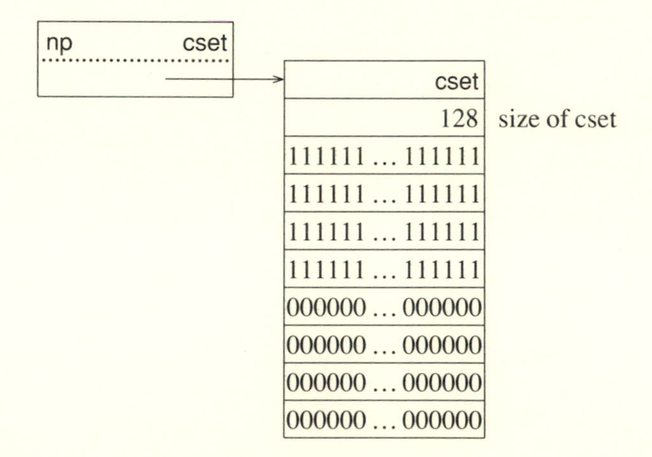

The first 128 bits are 1, since these are the bits that correspond to those in the ASCII character set.

The C structure for a cset block is

```
struct b_cset {                    /* cset block */
   word title;                     /*   T_Cset */
   word size;                      /*   size of cset */
   int bits[CsetSize];             /*   array of bits */
   };
```

where CsetSize is the number of words required to make up a total of 256 bits. CsetSize is 8 on a computer with 32-bit words and 16 on a computer with 16-bit words.

Cset operations are comparatively straightforward. The characters in a cset are represented by a bit vector that is divided into words to accommodate conventional computer architectures. For example, the C code cset complementation is

```
OpDcl(compl, 1, "~")
   {
   register int i, j;
   union block *bp;
   int *cs, csbuf[CsetSize];
   extern struct b_cset *alccset();

   blkreq((word)sizeof(struct b_cset));

   /*
    * Arg1 must be a cset.
    */
   if (cvcset(&Arg1, &cs, csbuf) == CvtFail)
      runerr(104, &Arg1);
```

```
/*
 * Allocate a new cset and then copy each cset word from Arg1
 * into the new cset words, complementing each bit.
 */
bp = (union block *)alccset(0);
for (i = 0; i < CsetSize; i++) {
    bp->cset.bits[i] = ~cs[i];
    }
  j = 0;
  for (i = 0; i < CsetSize * IntSize; i++) {
     if (Testb(i, bp->cset.bits))
        j++;
     }

bp->cset.size = j;
Arg0.dword = D_Cset;
BlkLoc(Arg0) = bp;
Return;
}
```

The macro Testb(b, c) tests bit b in cset c.

RETROSPECTIVE: The central role of strings in Icon and the nature of the operations performed on them leads to a representation of string data that is distinct from other data. The qualifier representation is particularly important in providing direct access to string length and in allowing the construction of substrings without the allocation of additional storage. The penalty paid is that a separate test must be performed to distinguish strings from all other kinds of values.

The ability to assign to subscripted strings causes serious implementation problems. The trapped-variable mechanism provides a solution, but it does so at considerable expense in the complexity of code in the run-time system as well as storage allocation for trapped-variable blocks. This expense is incurred even if assignment is not made to a subscripted string.

## EXERCISES

**5.1** What are the ramifications of Icon's use of the 256-bit ASCII character set, regardless of the ''native'' character set of the computer on which Icon is implemented?

**5.2** Catalog all the operations on strings in Icon and point out any that might cause special implementation problems. Indicate the aspects of strings and string operations in Icon that are the most important in terms of memory requirements and processing speed.

**5.3** List all the operations in Icon that require the allocation of space for the construction of strings.

**5.4** It has been suggested that it would be worth trying to avoid duplicate allocation of the same string by searching the string region for a newly created string to see if it already exists before allocating the space for it. Evaluate this proposal.

**5.5** Consider the following four expressions:

```
s1[i] := s2
s1[i+:1] := s2
a1[i] := a2
a1[i+:1] := a2
```

where s1 and s2 have string values and a1 and a2 have list values. Describe the essential differences between the string and list cases. Explain why these differences indicate flaws in language design. Suggest an alternative.

**5.6** The substring trapped-variable concept has the advantage of making it possible to handle all the contexts in which string-subscripting expressions can occur. It is expensive, however, in terms of storage utilization. Analyze the impact of this feature on the performance of ''typical'' Icon programs.

**5.7** Since the contexts in which most subscripting expressions occur can be determined, describe how to handle these without using trapped variables.

**5.8** If a subscripting expression is applied to a result that is not a variable, it is erroneous to use such an expression in an assignment context. In what situations can the translator detect this error? Are there any situations in which a subscripting expression is applied to a variable but in which the expression cannot be used in an assignment context?

**5.9** There are some potential advantages to unifying the keyword and substring trapped-variable mechanisms into a single mechanism in which all trapped variables would have pointers to functions for dereferencing and assignment. What are the disadvantages of such a unification?

**5.10** Presumably, it is unlikely for a programmer to have a constructive need for the polymorphic aspect of subscripting expressions. Or is it? If it is unlikely, provide a supporting argument. On the other hand, if there are situations in which this capability is useful, describe them and give examples.

**5.11** In some uses of map(s1,s2,s3), s1 and s2 remain fixed while s3 varies (Griswold 1980b). Devise a heuristic that takes advantage of such usage.

CHAPTER 6

# Lists

PERSPECTIVE: Most programming languages support some form of vector or array data type in which elements can be referenced by position. Icon's list data type fills this need, but it differs from similar types in many languages in that Icon lists are constructed during program execution instead of being declared during compilation. Therefore, the size of a list may not be known until run time.

Icon's lists are data objects. They can be assigned to variables and passed as arguments to functions. They are not copied when this is done; in fact, a value of type list is simply a descriptor that points to the structure that contains the list elements. These aspects of lists are shared by several other Icon data types and do not add anything new to the implementation. The attribute of lists that presents the most challenging implementation problem is their ability to grow and shrink by the use of stack and queue access mechanisms.

Lists present different faces to the programmer, depending on how they are used. They may be static vectors referenced by position or they may be dynamic, changing stacks or queues. It might seem that having a data structure with such apparently discordant access mechanisms would be awkward and undesirable. In practice, Icon's lists provide a remarkably flexible mechanism for dealing with many common programming problems. The two ways of manipulating lists are rarely intermixed. When both aspects are needed, they usually are needed at different times. For example, the number of elements needed in a list often is not known when the list is created. Such a list can be created with no elements, and the elements can be pushed onto it as they are produced. Once such a list has been constructed, it may be accessed by position with no further change in its size.

## 6.1 STRUCTURES FOR LISTS

The fusion of vector, stack, and queue organizations is reflected in the implementation of Icon by relatively complicated structures that are designed to provide a reasonable compromise between the conflicting requirements of the different access mechanisms.

A list consists of a fixed-size *list-header block*, which contains the usual title, the current size of the list (the number of elements in it), and descriptors that point to the first and last blocks on a doubly-linked chain of *list-element blocks* that contain the actual list elements. List-element blocks vary in size.

A list-element block contains the usual title, the size of the block in bytes, three words used to determine the locations of elements in the list-element block, and descriptors that point to the next and previous list-element blocks, if any. A null descriptor indicates the absence of a pointer to another list-element block. Following this data, there are slots for elements. Slots always contain valid descriptors, even if they are not used to hold list elements.

The structure declarations for list-header blocks and list-element blocks are

```
struct b_list {                    /* list-header block */
   word title;                     /*   T_List */
   word size;                      /*   current list size */
   struct descrip listhead;        /*   pointer to first list-element block */
   struct descrip listtail;        /*   pointer to last list-element block */
   };

struct b_lelem {                   /* list-element block */
   word title;                     /*   T_Lelem */
   word blksize;                   /*   size of block */
   word nslots;                    /*   total number of slots */
   word first;                     /*   index of first used slot */
   word nused;                     /*   number of used slots */
   struct descrip listprev;        /*   previous list-element block */
   struct descrip listnext;        /*   next list-element block */
   struct descrip lslots[1];       /*   array of slots */
   };
```

When a list is created, either by

```
list(n, x)
```

or by

```
[x1, x2, ..., xn]
```

there is only one list-element block. Other list-element blocks may be added to the chain as the result of pushs or puts.

List-element blocks have a minimum number of slots. This allows some expansion room for adding elements to lists, such as the empty list, that are small initially. The minimum number of slots is given by MinListSlots, which normally is eight. In the examples that follow, the value of MinListSlots is assumed to be four in order to keep the diagrams to a manageable size.

The code for the list function is

```
FncDcl(list, 2)
   {
   register word i, size;
   word nslots;
   register struct b_lelem *bp;
   register struct b_list *hp;
   extern struct b_list *alclist();
   extern struct b_lelem *alclstb();

   defshort(&Arg1, 0);                          /* size defaults to 0 */

   nslots = size = IntVal(Arg1);

   /*
    * Ensure that the size is positive and that the list-element block
    *  has at least MinListSlots slots.
    */
   if (size < 0)
      runerr(205, &Arg1);
   if (nslots < MinListSlots)
      nslots = MinListSlots;

   /*
    * Ensure space for a list-header block, and a list-element block
    * with nslots slots.
    */
   blkreq(sizeof(struct b_list) + sizeof(struct b_lelem) +
         nslots * sizeof(struct descrip));

   /*
    * Allocate the list-header block and a list-element block.
    *  Note that nslots is the number of slots in the list-element
    *  block while size is the number of elements in the list.
    */
   hp = alclist(size);
   bp = alclstb(nslots, (word)0, size);
   hp->listhead.dword = hp->listtail.dword = D_Lelem;
   BlkLoc(hp->listhead) = BlkLoc(hp->listtail) = (union block *)bp;

   /*
    * Initialize each slot.
    */
   for (i = 0; i < size; i++)
      bp->lslots[i] = Arg2;
```

```
/*
 * Return the new list.
 */
Arg0.dword = D_List;
BlkLoc(Arg0) = (union block *)hp;
Return;
}
```

The data structures produced for a list are illustrated by the result of evaluating

```
a := list(1, 4)
```

which produces a one-element list containing the value 4:

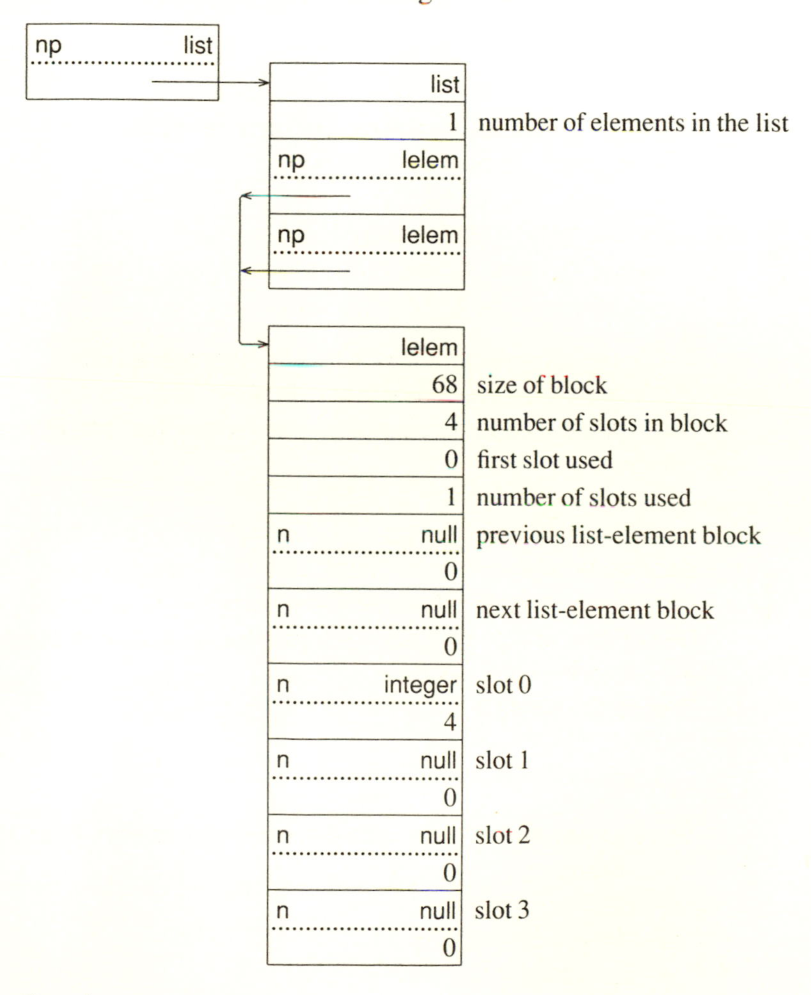


Data Structures for list(1, 4)

Note that there is only one list-element block and that the slot indexing in the block is zero-based. Unused slots contain null values that are logically inaccessible.

### 6.2 QUEUE AND STACK ACCESS

Elements in a list-element block are stored as a doubly-linked circular queue. If an element is added to the end of the list a, as in

```
put(a, 5)
```

the elements of the list are 4 and 5. The value is added to the ''end'' of the last list-element block, assuming there is an unused slot (as there is in this case). The code in put to do this is

```
/*
 * Point hp to the list-header block and bp to the last
 *  list-element block.
 */
hp = (struct b_list *)BlkLoc(Arg1);
bp = (struct b_lelem *)BlkLoc(hp->listtail);

/*
 * If the last list-element block is full, allocate a new
 *  list-element block, make it the first list-element block,
 *  and make it the next block of the former last list-element
 *  block.
 */
if (bp->nused >= bp->nslots) {
   bp = alclstb((word)MinListSlots, (word)0, (word)0);
   BlkLoc(hp->listtail)->lelem.listnext.dword = D_Lelem;
   BlkLoc(BlkLoc(hp->listtail)->lelem.listnext) = (union block *)bp;
   bp->listprev = hp->listtail;
   BlkLoc(hp->listtail) = (union block *)bp;
   }

/*
 * Set i to position of new last element and assign Arg2 to
 *  that element.
 */
i = bp->first + bp->nused;
if (i >= bp->nslots)
   i -= bp->nslots;
bp->lslots[i] = Arg2;
```

```
/*
 * Adjust block usage count and current list size.
 */
bp->nused++;
hp->size++;

/*
 * Return the list.
 */
Arg0 = Arg1;
Return;
}
```

The effect on the list-header block and list-element blocks in this example is

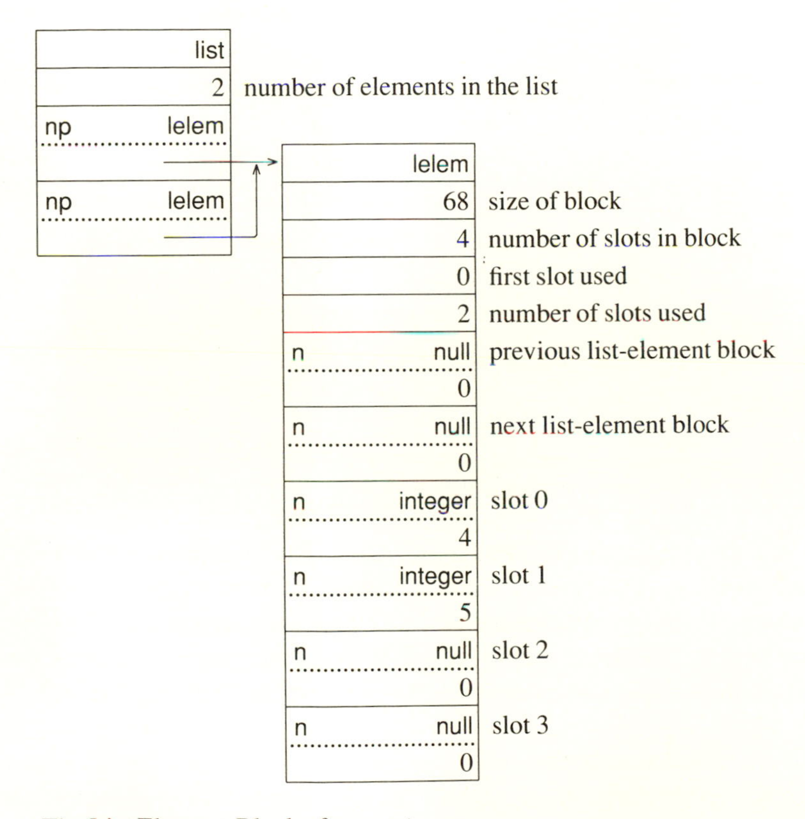


The List Element-Block after a put

Note that the increase in the number of elements in the list is reflected in the list-header block and in the number of slots used in the list-element block.

If an element is added to the beginning of a list, as in

```
push(a, 3)
```

the elements of the list are 3, 4, and 5. The new element is put at the "beginning" of the first list-element block. The result is

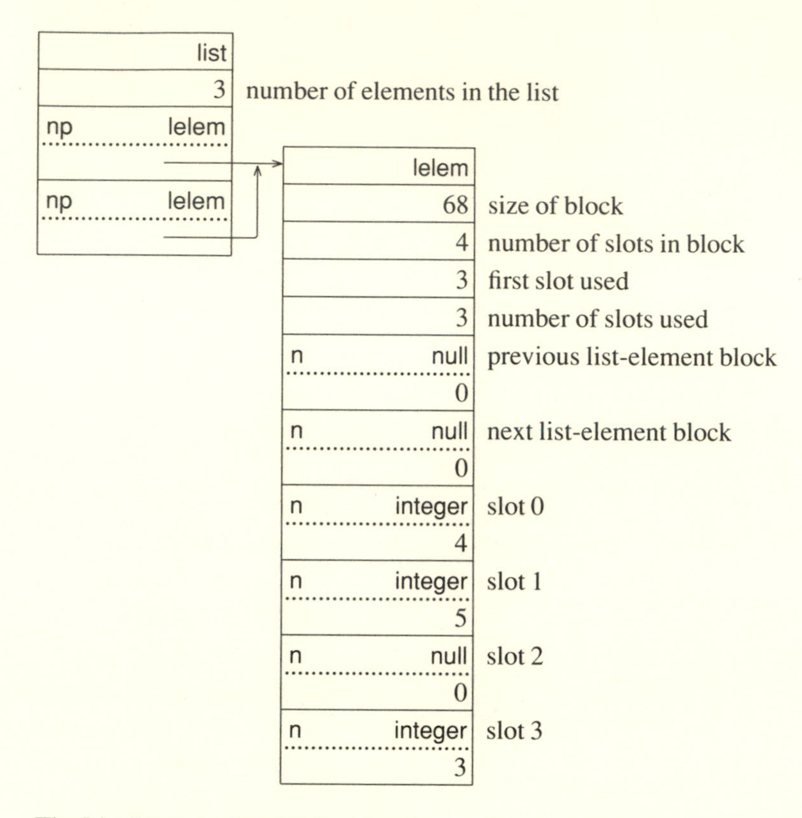


The List Element-Block after a push

Note that the "beginning," which is before the first physical slot in the list-element block, is the last physical slot. The locations of elements that are in a list-element block are determined by the three integers at the head of the list-element block. "Removal" of an element by a pop, get, or pull does not shorten the list-element block or overwrite the element; the element merely becomes inaccessible.

If an element is added to a list and no more slots are available in the appropriate list-element block, a new list-element block is allocated and linked in. For example, following evaluation of

```
push(a, 2)
push(a, 1)
```

the list elements are 1, 2, 3, 4, and 5. The resulting structures are

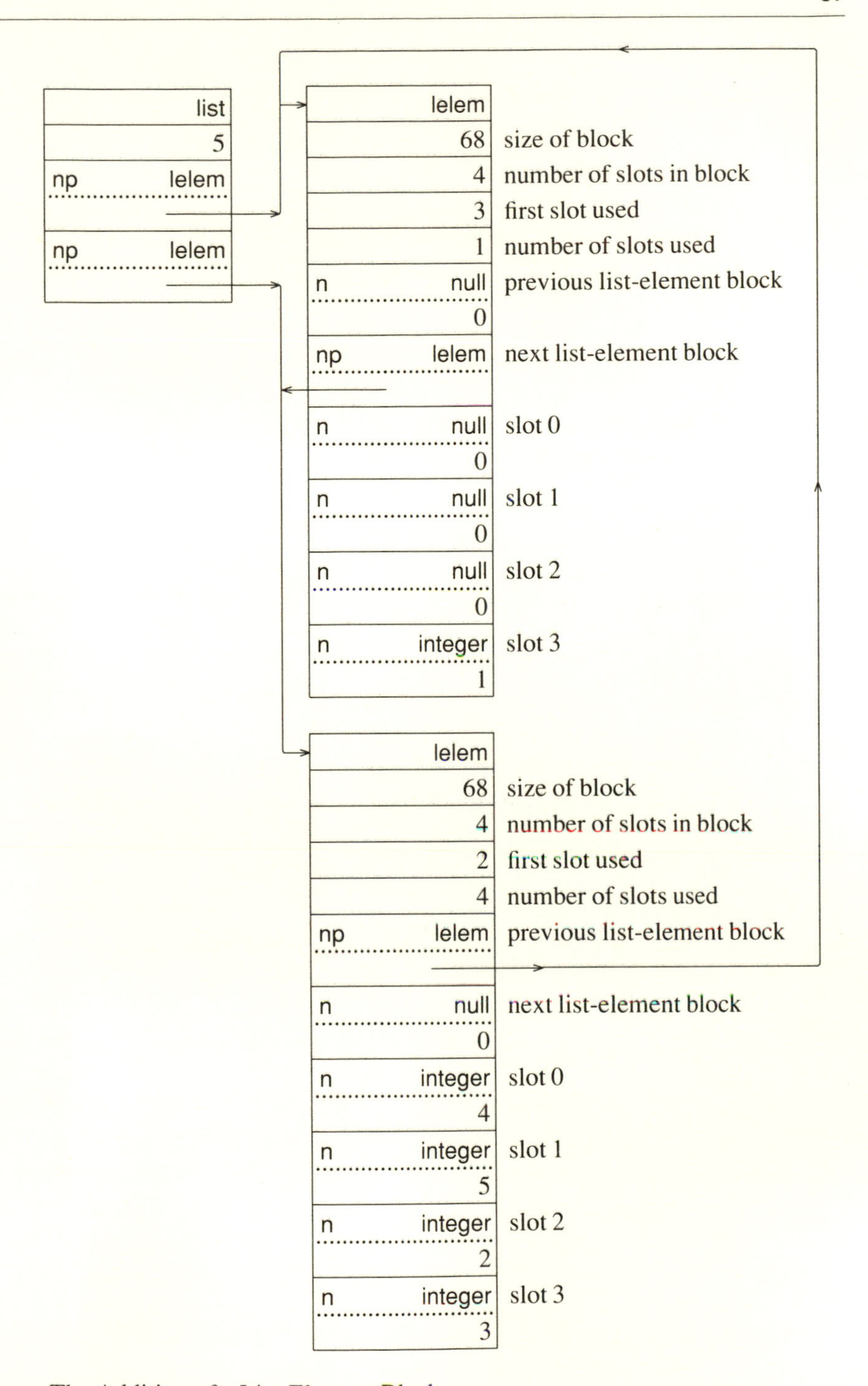


The Addition of a List-Element Block

As elements are removed from a list by put (which is synonymous with get) or pull, the indices in the appropriate list-element block are adjusted. The code for pop is

```
FncDcl(pop, 1)
   {
   register word i;
   register struct b_list *hp;
   register struct b_lelem *bp;
   extern struct b_lelem *alclstb();

   /*
    * Arg1 must be a list.
    */
   if (Arg1.dword != D_List)
      runerr(108, &Arg1);

   /*
    * Fail if the list is empty.
    */
   hp = (struct b_list *)BlkLoc(Arg1);
   if (hp->size <= 0)
      Fail;

   /*
    * Point bp to the first list-element block. If the first block has
    *  no slots in use, point bp at the next list-element block.
    */
   bp = (struct b_lelem *)BlkLoc(hp->listhead);
   if (bp->nused <= 0) {
      bp = (struct b_lelem *)BlkLoc(bp->listnext);
      BlkLoc(hp->listhead) = (union block *)bp;
      bp->listprev = nulldesc;
      }
```

```
/*
 * Locate first element and assign it to Arg0 for return.
 */
i = bp->first;
Arg0 = bp->lslots[i];

/*
 * Set bp->first to new first element, or 0 if the block is now
 *  empty. Decrement the usage count for the block and the size
 *  of the list.
 */
if (++i >= bp->nslots)
   i = 0;
bp->first = i;
bp->nused--;
hp->size--;
Return;
}
```

Thus, as a result of

```
pop(a)
```

the list elements are 2, 3, 4, and 5. The resulting structures are

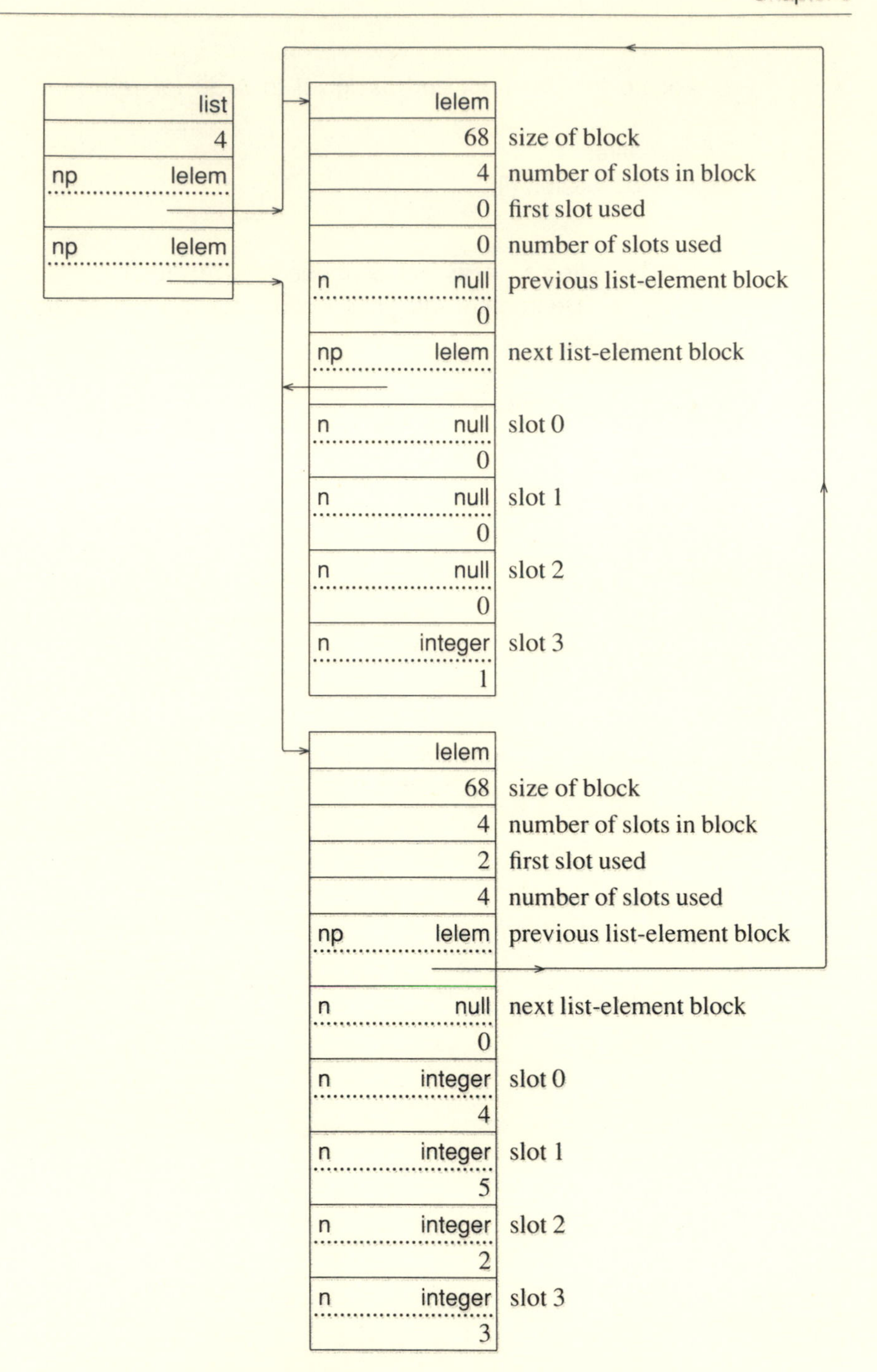


The Result of Removing Elements from a List-Element Block

Note that the first list-element block is still linked in the chain, even though it no longer contains any elements that are logically accessible. A list-element block is not removed from the chain when it becomes empty. It is removed only when an element is removed from a list that already has an empty list-element block. Thus, there is always at least one list-element block on the chain, even if the list is empty. Aside from simplifying the access to list-element blocks from the list-header block, this strategy avoids repeated allocation in the case that pop/push pairs occur at the boundary of two list-element blocks.

Continuing the previous example,

```
pop(a)
```

leaves the list elements 3, 4, and 5. The empty list-element block is removed from the chain:

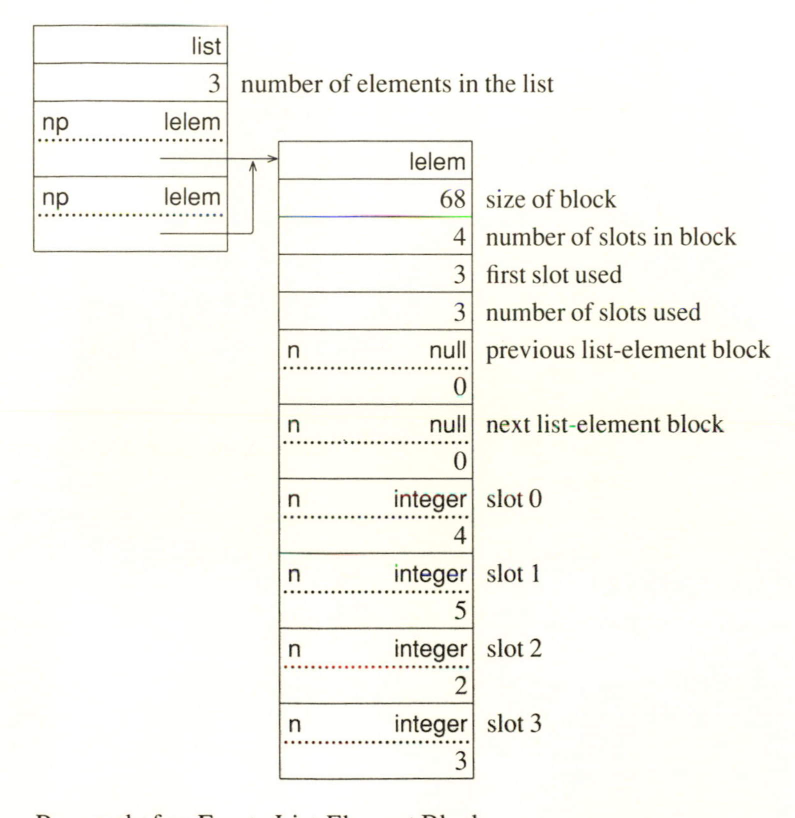


Removal of an Empty List-Element Block

Note that the value 2 is still physically in the list-element block, although it is logically inaccessible.

## 6.3 POSITIONAL ACCESS

A positional reference of the form a[i] requires locating the correct list-element block. Out-of-range references can be determined by examining the list-header block. If the list has several list-element blocks, this involves linking through the list-element blocks, while keeping track of the count of elements in each block until the appropriate one is reached. The result of evaluating a[i] is a variable that points to the appropriate slot.

The portion of the subscripting code that handles lists is

```
switch (Type(Arg1)) {
   case T_List:
      /*
       * Make sure that Arg2 is an integer and that the
       *  subscript is in range.
       */
      if (cvint(&Arg2, &I1) == CvtFail)
         runerr(101, &Arg2);
      i = cvpos(I1, BlkLoc(Arg1)->list.size);
      if (i == 0 || i > BlkLoc(Arg1)->list.size)
         Fail;

      /*
       * Locate the list-element block containing the desired
       *  element.
       */
      bp = BlkLoc(BlkLoc(Arg1)->list.listhead);
      j = 1;
      while (i >= j + bp->lelem.nused) {
         j += bp->lelem.nused;
         bp = BlkLoc(bp->lelem.listnext);
         }
```

```
/*
 * Locate the desired element and return a pointer to it.
 */
i += bp->lelem.first - j;
if (i >= bp->lelem.nslots)
   i -= bp->lelem.nslots;
dp = &bp->lelem.lslots[i];
Arg0.dword = D_Var + ((int *)dp - (int *)bp);
VarLoc(Arg0) = dp;
Return;
```

For the preceding example, a[3] produces a variable that points to the descriptor for the value 5:

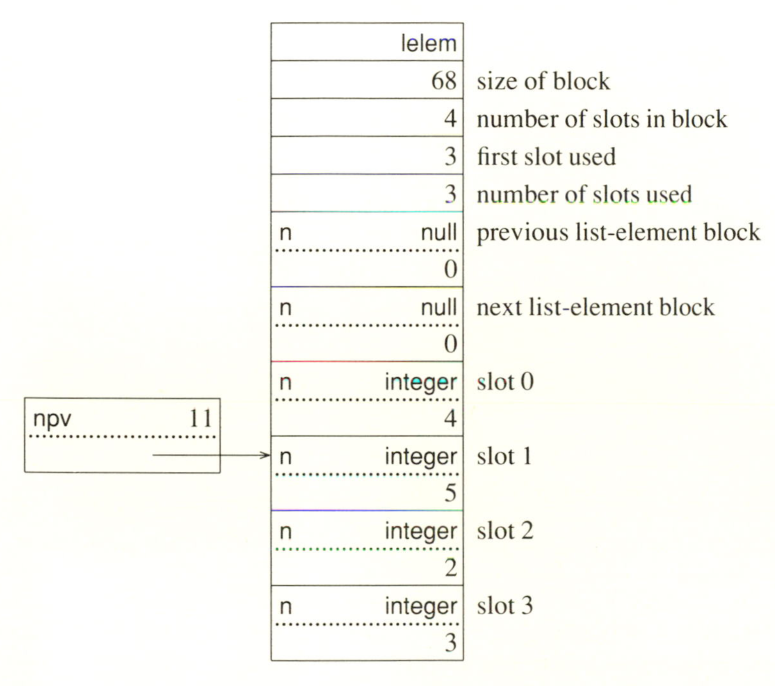


Referencing a List Element

Note the offset of eleven words in the d-word of the variable. This is present so that the title of the block to which the variable points can be located in case there is a garbage collection. See Chapter 11 for details.

RETROSPECTIVE: The structures used for implementing lists are relatively complicated, but they provide a reasonable compromise, both in the utilization of storage and access speed, that accommodates different access mechanisms.

Using a chain of list-element blocks allows lists to grow in size without limit. From the viewpoint of positional access, this amounts to segmentation. This segmentation only occurs, however, when elements are added to a list.

The use of circular queues within list-element blocks allows elements to be removed and added without wasting space.

## EXERCISES

**6.1** Diagram the structures that result from the evaluation of the following expressions:

```
graph := ["a",,]
graph[2] := graph[3] := graph
```

**6.2** How much space does an empty list occupy?

**6.3** The portions of the structures for a list that are not occupied by elements of the list constitute overhead. Calculate the percentage of overhead in the following lists. Assume that the minimum number of slots in a list-element block is eight.

```
a := []

a := [1,2]

a := [1,2,3,4,5]

a := list(100)

a := []; every put(a,1 to 100)
```

How do these figures vary as a function of the minimum number of slots in a list-element block?

**6.4** What are the implications of not ''zeroing'' list elements when they are logically removed by a pop, get, or pull?

**6.5** When a list-element block is unlinked as the result of a pop, get, or pull, are the elements in it really inaccessible to the source program?

**6.6** There is considerable overhead involved in the implementation of lists to support both positional access and stack and queue access mechanisms. Suppose the language were changed so that stack and queue access mechanisms applied only to lists that were initially empty. What would the likely impact be on existing Icon programs? How could the implementation take advantage of this change?

**6.7** As elements are added to lists, more list-element blocks are added and they tend to become ''fragmented.'' Is it feasible to reorganize such lists, combining the elements in many list-element blocks into one large block? If so, when and how could this be done?

**6.8** A suggested alternative to maintaining a chain of list-element blocks is to allocate a larger block when space is needed and copy elements from the previous block into it. Criticize this proposal.

**6.9** Suppose it were possible to insert elements in the middle of lists, rather than only at the ends. How might this feature be implemented?

CHAPTER 7

# Sets and Tables

PERSPECTIVE: Sets and tables are data aggregates that are very useful for a number of common programming tasks. Nevertheless, few programming languages support these data types, with the notable exceptions of Sail (Reiser 1976) and SETL (Dewar, Schonberg, and Schwartz 1981). There are many reasons why these obviously useful data types are not found in most programming languages, but perceived implementation problems certainly rank high among them. If only for this reason, their implementation in Icon is worth studying.

Historically, tables in Icon were inherited from SNOBOL4 and SL5. Sets came later, as an extension to Icon, and were designed and implemented as a class project. Although sets were a late addition to Icon, they are simpler than tables. Nonetheless, they present many of the same implementation problems that tables do. Consequently, sets are considered here first.

Sets and the operations on them support the familiar mathematical concepts of finite sets: membership, the insertion and deletion of members, and the operations of union, intersection, and difference. What is interesting about a set in Icon is that it can contain members of any data type. This is certainly a case where heterogeneity significantly increases the usefulness of a data aggregate without adding to the difficulty of the implementation, *per se*.

The ability of a set to grow and shrink in size influences the implementation significantly. Efficient access to members of a set, which is needed for testing membership as well as the addition and deletion of members, is an important consideration, since sets can be arbitrarily large.

Tables have more structure than sets. Abstractly, a table is a set of pairs that represents a many-to-one relationship—a function. In this sense, the default value of a table provides an extension of the partial function represented by the entry and assigned value pairs to a complete function over all possible entry values. Programmers, however, tend to view tables in a more restricted way, using them to tabulate the attributes of a set of values of interest. In fact, before sets were added to Icon, tables were often used to simulate sets by associating a specific assigned value with membership.

## 7.1 SETS

### 7.1.1 Data Organization for Sets

Hash lookup and linked lists are used to provide an efficient way of locating set members. For every set there is a set-header block that contains a word for the number of members in the set and slots that serve as heads for (possibly empty) linked lists of set-element blocks. The number of slots is an implementation parameter. There are thirty-seven slots in table-header blocks on computers with large address spaces but only thirteen slots on computers with small address spaces.

The structure for an empty set, produced by

```
s := set([])
```

is

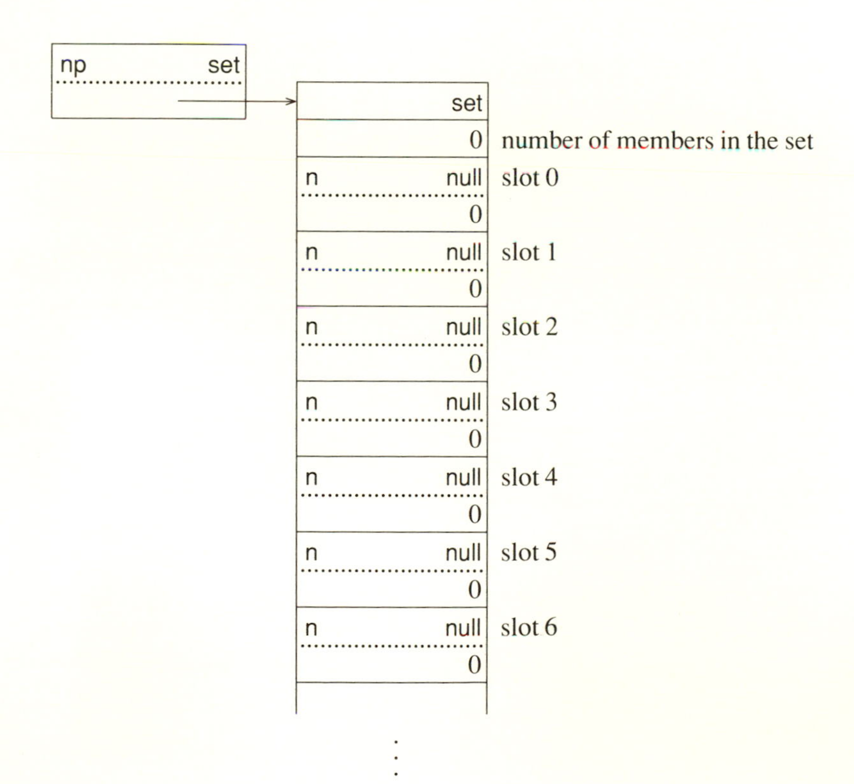

Each member of a set is contained in a separate set-element block. When a value is looked up in a set (for example, to add a new member), a hash number is computed from this value. The absolute value of the remainder resulting from dividing the hash number by the number of slots is used to select a slot.

Each set-element block contains a descriptor for its value, the corresponding hash number, and a pointer to the next set-element block, if any, on the linked list. For example, the set-element block for the integer 39 is:

| block | |
|---|---|
| selem | |
| 39 | hash number |
| n null | next set-element block |
| 0 | |
| n integer | member value |
| 39 | |

As illustrated by this figure, the hash number for an integer is just the value of the integer. This member goes in slot 2 on computers with large address spaces, since its remainder on division by the number of slots is two. Hash computation is discussed in detail in Sec. 7.3.

The structures for the set

```
s := set([39, 2])
```

are

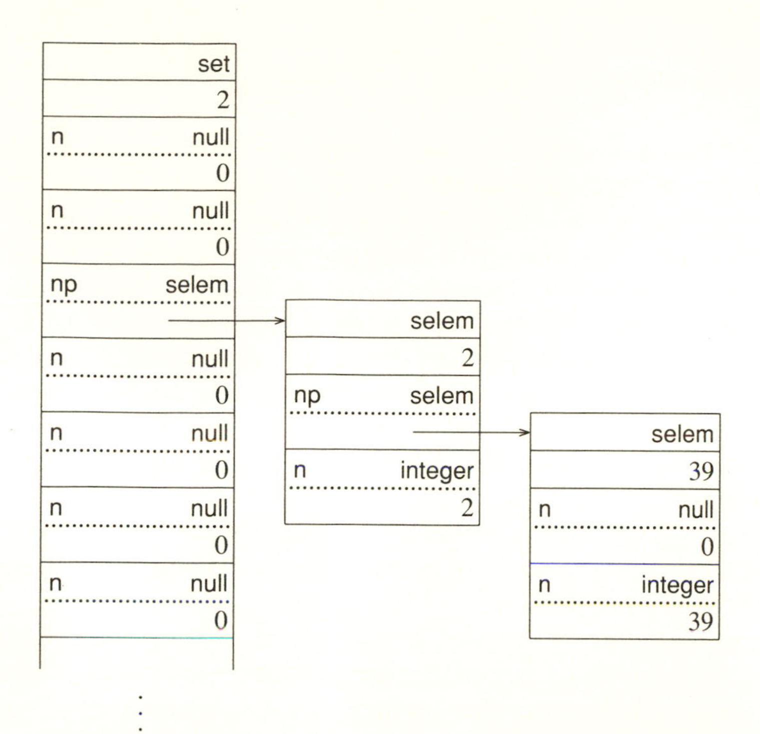


This example was chosen for illustration, since both 2 and 39 go in slot 2.

In searching the list, the hash number of the value being looked up is compared with the hash numbers in the set-element blocks. If a match is found, the value in the set-element block may or may not be the same as the value being looked up, since collisions in the hash computation are unavoidable. Thus, if the hash numbers are the same, it is necessary to determine whether or not their values are equivalent. The comparison that is used is the same one that is used by the source-language operation x === y.

To improve the performance of the lookup process, the set-element blocks in each linked list are ordered by their hash numbers. When a linked list of set-element blocks is examined, the search stops if a hash number of an element on the list is greater than the hash number of the value being looked up.

If the value is not found and the lookup is being performed to insert a new member, a set-element block for the new member is created and linked into the list at that point. For example,

```
insert(s, -39)
```

inserts a set-element block for –39 at the head of the list in slot 2, since its hash value is –39. The word in the set-header block that contains the number of members is incremented to reflect the insertion.

### 7.1.2 Set Operations

The set operations of union, intersection, and difference all produce new sets and do not modify their arguments.

In the case of union, a copy of the larger set is made first to provide the basis for the union. This involves not only copying the set-header block but also all of its set-element blocks. These are linked together as in the original set, and no lookup is required. After this copy is made, each member of the set for the other argument is inserted in the copy, using the same technique that is used in insert. The larger set is copied, since copying does not require lookup and the possible comparison of values that insertion does. The insertion of a member from the second set may take longer, however, since the linked lists in the copy may be longer.

In the case of intersection, a copy of the smaller argument set is made, omitting any of its members that are not in the larger set. As with union, this strategy is designed to minimize the number of lookups.

For the difference of two sets, a copy of the first argument set is made, adding only elements that are not in the second argument. This involves looking up all members in the first argument set in the second argument set.

## 7.2 TABLES

### 7.2.1 Data Organization for Tables

The implementation of tables is similar to the implementation of sets, with a header block containing slots for elements ordered by hash numbers. A table-header block contains an extra descriptor for the default assigned value.

An empty table with the default assigned value 0 is produced by

```
t := table(0)
```

The structure of the table-header block is

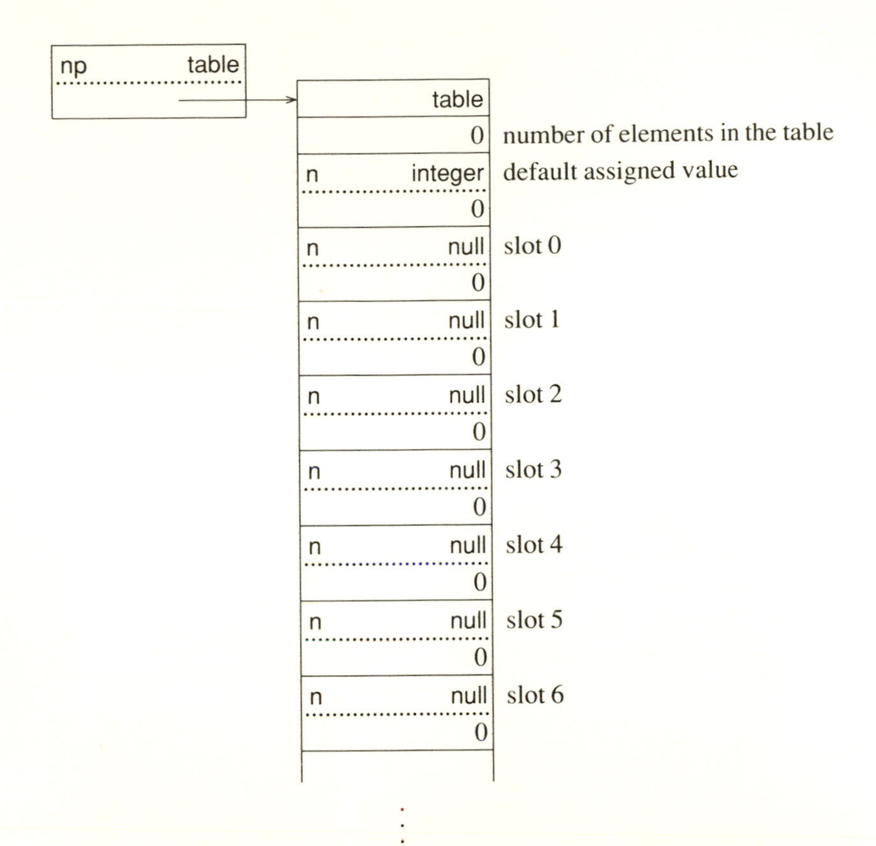


Table lookup is more complicated than set lookup, since table elements contain both an entry value and an assigned value. Furthermore, table elements can be referenced by variables. A new table element is created as a byproduct of assignment to a table reference with an entry value that is not in the table.

The result of evaluating an assignment expression such as

```
t[39] := 1
```

illustrates the structure of a table-element block:

| | | |
|---|---|---|
| | telem | |
| | 39 | hash number |
| n | null | next table-element block |
| | 0 | |
| n | integer | entry value |
| | 39 | |
| n | integer | assigned value |
| | 1 | |

In the case of a table reference such as t[x], the hash number for the entry value x is used to select a slot, and the corresponding list is searched for a table-element block that contains the same entry value. As in the case of sets, comparison is first made using hash numbers; values are compared only if their hash numbers are the same.

If a table-element block with a matching entry value is found, a variable that points to the corresponding assigned value is produced. For example, if 39 is in t as illustrated previously, t[39] produces

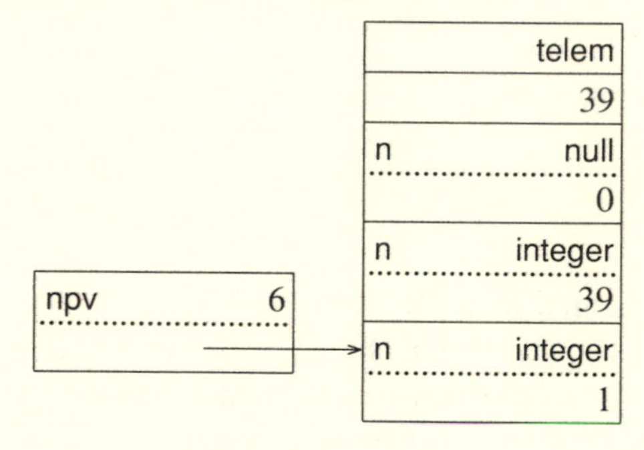


If this variable is dereferenced, as in

```
write(t[39])
```

the value 1 is written. On the other hand, if an assignment is made to this variable, as in

```
t[39] +:= 1
```

the assigned value in the table-element block is changed:

| | telem |
|---|---:|
| | 39 |
| n | null |
| | 0 |
| n | integer |
| | 39 |
| n | integer |
| | 2 |

If a table element with a matching entry value is not found, the situation is very similar to that in a subscripted string: the operation to be performed depends on whether the table reference is used in a dereferencing or assignment context. In a dereferencing context, the default value for the table is produced, while in an assignment context, a new element is added to the table.

The approach taken is similar to that for subscripted strings: a trapped variable is created. As with substring trapped variables, table-element trapped variables contain the information that is necessary to carry out the required computation for either dereferencing or assignment.

Suppose, for example, that the entry value 36 is not in the table t. Then t[36] produces the following result:

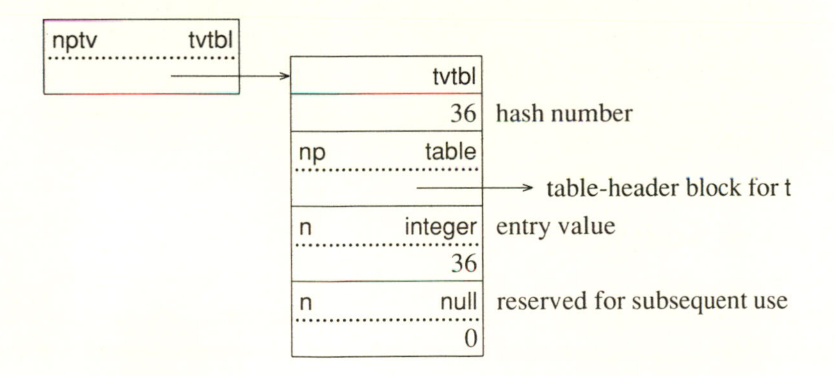


Note that the size of a table-element trapped-variable block is the same as the size of a table-element block. The last descriptor in the table-element trapped-variable block is reserved for subsequent use, as described below.

If this trapped variable is dereferenced, as in

```
write(t[36])
```

the default assigned value, 0, which is in the table-header block for t, is produced. Unfortunately, the situation is not always this simple. It is possible for elements to be inserted in a table between the time the table-element trapped-variable block is created and the time it is dereferenced. An example is

```
write(t[36], t[36] := 2)
```

Since functions do not dereference their arguments until all the arguments have been evaluated, the result of dereferencing the first argument of write should be 2, not 0. In order to handle such cases, when a table-element trapped variable is dereferenced, its linked list in the table must be searched again to determine whether to return the assigned value of a newly inserted element or to return the default value.

If an assignment is made to the table reference, as in

```
t[36] +:= 1
```

the table-element trapped-variable block is converted to a table-element block with the assigned value stored in the reserved descriptor of the table-element trapped-variable block. The table-element block is then linked in the appropriate place. Note that the structures of table-element blocks and table-element trapped-variable blocks are the same, allowing this conversion without allocating a new table-element block.

It then is necessary to search the linked list for its slot again to determine the place to insert the table-element block. As in the case of dereferencing, elements may have been inserted in the table between the time the table-element trapped variable was created and the time a value is assigned to it. Normally, no matching entry is found, and the table-element trapped-variable block, transformed into a table-element block, is inserted with the new assigned value. If a matching entry is found, its assigned value is simply changed, and the block is discarded.

Note that reference to a value that is not in a table requires only one computation of its hash value, but two lookups are required in the linked list of table-element blocks for its slot.

## 7.3 HASHING FUNCTIONS

Ideally, a hash computation should produce a different result for every different value to which it is applied, and the distribution of the remainder on division by the number of slots should be uniform. Even approaching this ideal requires an impractical amount of computation and space. In practice, it is desirable to have a fast computation that produces few collisions.

The subject of hash computation has been studied extensively and there is a substantial body of knowledge concerning useful techniques (Knuth 1973, pp. 506-549). For example, it is known that the number of slots should be a prime that is not close to a power of two. This consideration motivated the choices of 37 and 13 for computers with large and small address spaces, respectively. In general, there is a trade-off between faster lookup, on the average, and more storage overhead.

In most situations in which hashing techniques are used, all the values for which hash computations are performed are strings. In Icon, however, any kind of value can be the member of a set or the entry value in a table. The hash computation must, therefore, apply to any type of value. The support routine for computing hash numbers is

```
word hash(dp)
struct descrip *dp;
   {
   word i;
   double r;
   register word j;
   register char *s;

   if (Qual(*dp)) {

      /*
       * Compute the hash value for the string by summing the value
       *  of all the characters (to a maximum of 10) plus the length.
       */
      i = 0;
      s = StrLoc(*dp);
      j = StrLen(*dp);
      for (j = (j <= 10) ? j : 10 ; j > 0; j--)
         i += *s++ & 0377;
      i += StrLen(*dp) & 0377;
      }
   else {
      switch (Type(*dp)) {
         /*
          * The hash value for numeric types is the bit-string
          *  representation of the value.
          */

         case T_Integer:
            i = IntVal(*dp);
            break;

         case T_Long:
            i = BlkLoc(*dp)->longint.intval;
            break;
```

```
            case T_Real:
                GetReal(dp, r);
                i = r;
                break;

            case T_Cset:
                /*
                 * Compute the hash value for a cset by performing the
                 *  exclusive-or of the words in the bit array.
                 */
                i = 0;
                for (j = 0; j < CsetSize; j++)
                    i ^= BlkLoc(*dp)->cset.bits[j];
                break;

            default:
                /*
                 * For other types, use the type code as the hash
                 *  value.
                 */
                i = Type(*dp);
                break;
            }
        }

    return i;
    }
```

To hash a string, its characters are added together as integers. At most ten characters are used, since strings can be very long and adding many characters does not improve the hashing sufficiently to justify the time spent in the computation. The maximum of ten is, however, *ad hoc*. To provide a measure of discrimination between strings with the same initial substring, the length of the string is added to the sum of the characters. This technique for hashing strings is not sophisticated, and others that produce better hashing results are known. However, the computation is simple, easy to write in C, and works well in practice.

For a numeric type, the hash value is simply the number. In the case of a cset, the words containing the bits for the cset are combined using the exclusive-or operation, and the result is cast as an integer.

The remaining data types pose an interesting problem. Hash computation must be based on attributes of a value that are invariant with time. Some types, such as files, have such attributes. On the other hand, there is no time-invariant attribute that distinguishes one list from another. The size of a list may change, the elements in it may change, and even its location in memory may change as the result of garbage collection. For a list, its only time-invariant attribute is its type.

This presents a dilemma—the type of such a value can be used as its hash number, but if that is done, all values of that type are in the same slot and have the same hash number. Lookup for these values degenerates to a linear search. The alternative is to add some time-invariant attribute, such as a serial number, to these values. This would increase the size of every such value, however.

Hash computation in Icon resolves this problem in favor of simplicity. Lists and similar values are hashed according to their type codes. Part of the rationale for this choice is that it is uncommon, in practice, to have sets of lists, tables of sets, and so forth. On balance, it probably is not worth adding the space overhead for every such value just to improve the performance of only a few programs.

RETROSPECTIVE: Few programming languages support sets or tables; fewer support them with Icon's generality. The implementation of sets and tables provides a clear focus on the generality of descriptors and the uniformity with which different kinds of data are treated in Icon.

Since sets and tables may be very large, efficient lookup is an important concern. The hashing and chaining technique used is only one of many possibilities. However, there must be a mechanism for determining the equivalence of values independent of the structure in which they are stored.

The fact that elements in tables are accessed by subscripting expressions introduces several complexities. In particular, the fact that the contents of the table that is subscripted may change between the time the subscripting expression is evaluated and the time it is dereferenced or assigned to introduces the necessity of two lookups for every table reference.

Hashing a variety of different types of data raises interesting issues. The hashing techniques used by Icon are not sophisticated and there is considerable room for improvement. The trade-offs involved are difficult to evaluate, however.

## EXERCISES

**7.1** Contrast sets and csets with respect to their implementation, their usefulness in programming, and the efficiency of operations on them.

**7.2** Give an example of a situation in which the heterogeneity of sets is useful in programming.

**7.3** How much space does an empty set occupy?

**7.4** Diagram the structures resulting from the evaluation of the following expressions:

```
t := table()
t[t] := t
```

**7.5** There are many sophisticated data structures that are designed to ensure efficient lookup in data aggregates like sets and tables (Gonnet 1984). Consider the importance of speed of lookup in sets and tables in Icon and the advantages that these more sophisticated data structures might supply.

**7.6** Some of the more sophisticated data structures mentioned in the preceding exercise have been tried experimentally in Icon and either have introduced unexpected implementation problems or have not provided a significant improvement in performance. What are possible reasons for these disappointing results?

**7.7** Icon goes to a lot of trouble to avoid adding table-element blocks to a table unless an assignment is made to them. Suppose a table-element block were simply added when a reference was made to an entry value that is not in the table.

- How would this simplify the implementation?
- What positive and negative consequences could this change have on the running speed and space required during program execution?
- Give examples of types of programs for which the change would have positive and negative effects on performance, respectively.
- Would this change be transparent to the Icon programmer, not counting possible time and space differences?

**7.8** There is space in a table-element trapped-variable block to put the default value for the table. Why is this not done?

**7.9** What is the consequence of evaluating the following expressions?

```
t := table(0)
t[37] := 2
write(t[37], t := table(1))
```

What would happen if the last line given previously were

```
write(t[37], t := list(100, 3))
```

or

```
write(t[37], t := "hello")
```

**7.10** Give examples of different strings that have the same hash numbers.

**7.11** Design a method for hashing strings that produces a better distribution than the current one.

**7.12** What attribute of a table is time-invariant?

**7.13** What kinds of symptoms might result from a hashing computation based on an attribute of a value that is not time-invariant?

CHAPTER 8

# The Interpreter

PERSPECTIVE: The interpreter provides a software realization of Icon's virtual machine. This machine is stack-based. The basic units on which the Icon virtual machine operates are descriptors. The instructions for the virtual machine consist of operations that manipulate the stack, call C functions that carry out the built-in operations of Icon, and manage the flow of control. The Icon interpreter executes these virtual machine instructions. It consists of a loop in which a virtual machine instruction is fetched and control is transferred to a section of code to perform the corresponding operation.

## 8.1 STACK-BASED EVALUATION

Virtual machine instructions typically push and pop data on the interpreter stack. The interpreter stack, which is distinct from the stack used for calls of C functions, is an array of words. The variable sp points to the last word pushed on the interpreter stack. Pushing increments sp, while popping decrements it. When the interpreter executes code that corresponds to a built-in operation in Icon, it pushes descriptors for the arguments on the interpreter stack and calls a C function corresponding to that operation with a pointer to the place on the interpreter stack where the arguments begin. A null descriptor is pushed first to serve as a "zeroth" argument (Arg0) that receives, by convention, the result of the computation and becomes the top descriptor on the stack when the C function returns. On a more conventional virtual machine, the result of the computation would be pushed on the stack, instead of being returned in an argument. The latter method is more convenient in Icon.

To illustrate this basic mechanism, consider the expression

```
?10
```

which produces a randomly selected integer between 1 and 10, inclusive. The corresponding virtual machine instructions are

```
pnull            # push null descriptor for the result
int       10     # push descriptor for the integer 10
random           # compute random value
```

The instructions pnull and int operate directly on the stack. The instruction random calls a C function that computes random values.

The pnull instruction pushes a null descriptor:

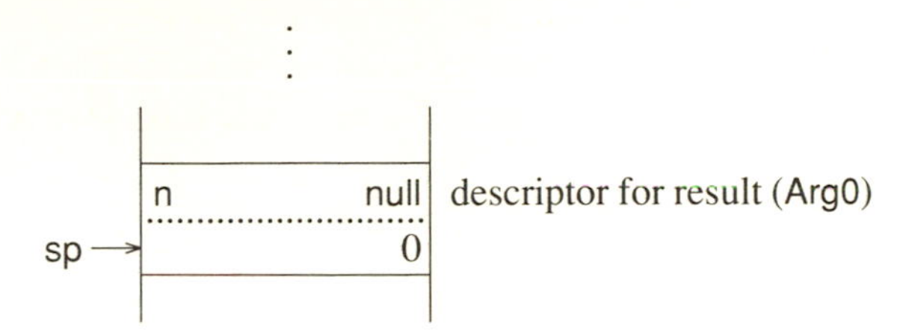


The int instruction pushes a descriptor for the integer 10:

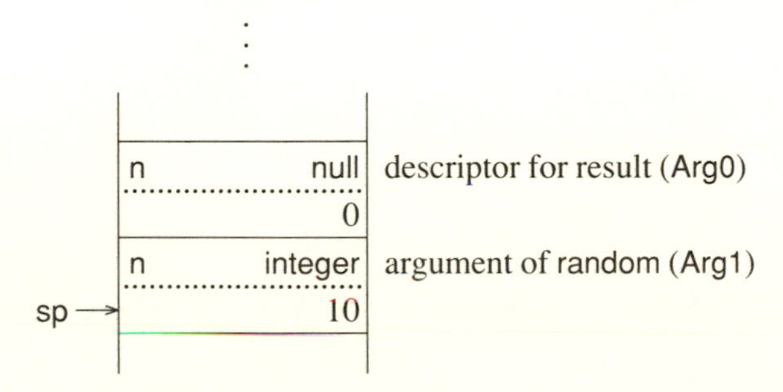


Suppose that the C function for random computes 3. It replaces the null value of Arg0 by a descriptor for the integer 3. When it returns, sp is set to point to Arg0 and the situation is

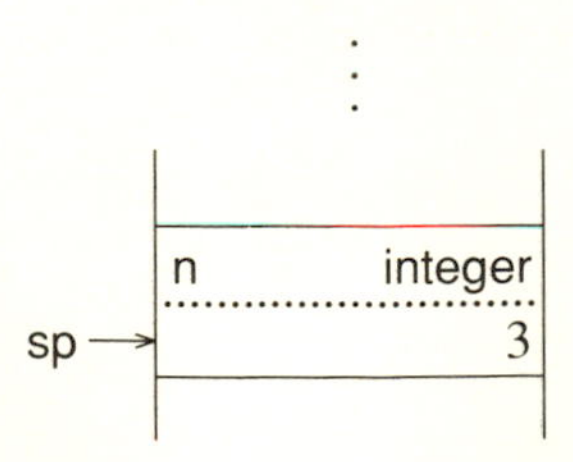


## 8.2 VIRTUAL MACHINE INSTRUCTIONS

The various aspects of expressions that appear in Icon source-language programs are reflected, directly or indirectly, in the instruction set for the Icon virtual machine. References to constants (literals) and identifiers have direct correspondences in the instruction set of the virtual machine. There is a virtual machine instruction for each source-language operator. This is possible, since the meaning of an operation is fixed and cannot be changed during program execution. The meaning of a function call, however, cannot be determined until it is evaluated, and there is a single virtual machine instruction for function invocation. The invocation of functions is described in detail in Chapter 10.

There are several virtual machine instructions related to control structures and the unique aspects of expression evaluation in Icon. These are discussed in the next two chapters. A complete list of virtual machine instructions is given in Appendix B.

### 8.2.1 Constants

Four kinds of data can be represented literally in Icon programs: integers, strings, csets, and real numbers. The four corresponding virtual machine instructions are

```
int     n      # integer n
str     n, a   # string of length n at address a
cset    a      # cset block at address a
real    a      # real block at address a
```

The values of integer literals appear as arguments of int instructions. In the case of strings, the two arguments give its length and the address of its first character. The string itself is constructed by the linker and is loaded into memory from the icode file. For csets and real numbers, the linker constructs blocks, which are also loaded from the icode file. These blocks are identical in format to blocks that are constructed during program execution.

The virtual machine instructions str, cset, and real push appropriate descriptors to reference the data as it appears in the icode. For example, the virtual machine instructions for

```
?"aeiou"
```

are

```
pnull
str     5, a
random
```

where a is the address of the string "aeiou". The pnull instruction pushes a null descriptor as in the previous example:

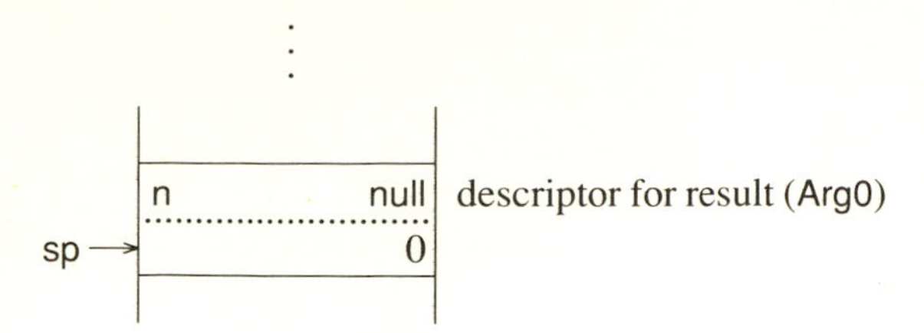


The str instruction constructs a descriptor for the string "aeiou":

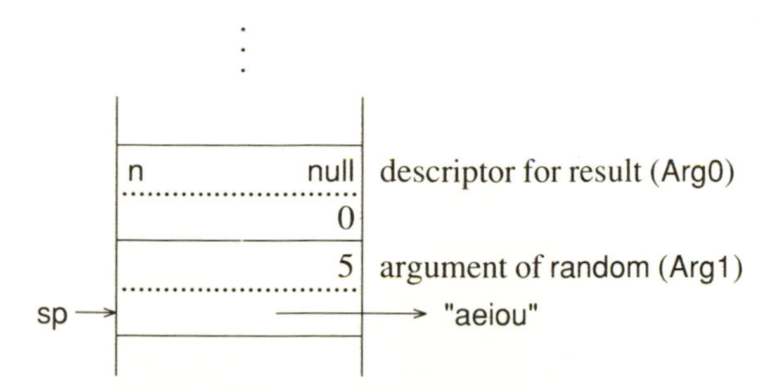


If random produces the string "o", this string replaces the null descriptor and the stack becomes

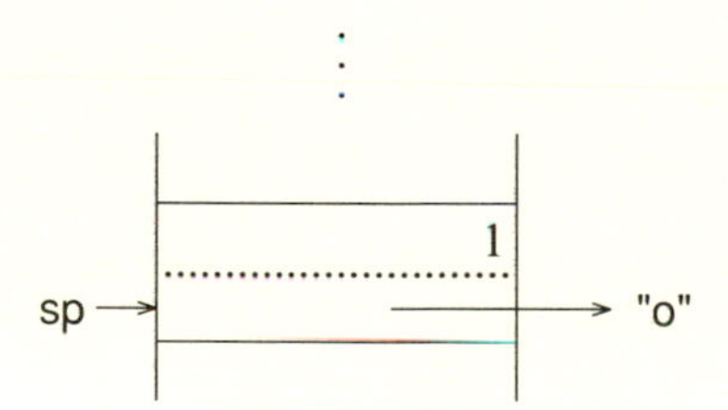


### 8.2.2 Identifiers

From the viewpoint of the interpreter, there are four kinds of identifiers: global identifiers, static identifiers, local identifiers, and arguments. The values of global and static identifiers are in arrays of descriptors at fixed locations in memory. The values of local identifiers and arguments, on the other hand, are kept on the stack as part of the information associated with a procedure call.

The values of the arguments in the call of a procedure are pushed on the stack as the result of the evaluation of expressions prior to the invocation of the procedure. The initial null values for local identifiers are pushed on the stack when the procedure is called.

Consider the following procedure declaration:

```
procedure p(x, y)
   local z, i, j
   j := 1
   z := x
     .
     .
     .
end
```

In the evaluation of a call to this procedure, such as

```
p(10, 20)
```

the stack is in the following state prior to the evaluation of the first expression in p:

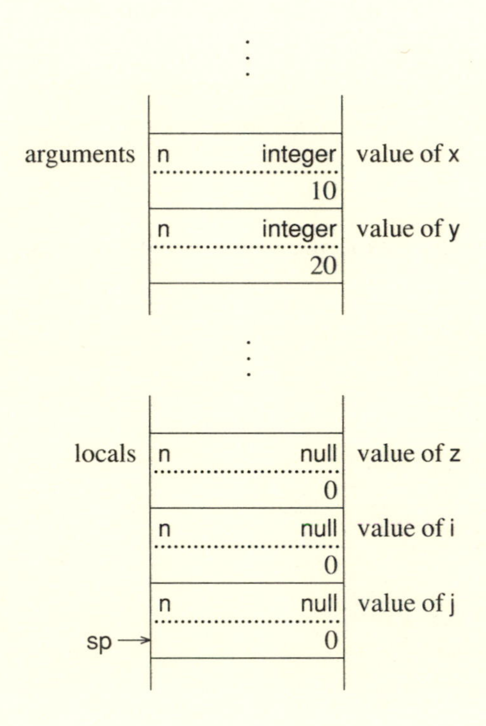

The portion of the stack between the arguments and local identifiers is fixed in size and contains information that is saved when a procedure is called. This information is described in Chapter 10.

There are four virtual machine instructions for constructing variable descriptors:

```
global    n
static    n
arg       n
local     n
```

Identifiers of each kind are numbered starting at zero. Consequently,

```
arg       0
```

pushes a variable descriptor for the first argument. In each case, the descriptor that is pushed on the stack is a variable that points to the descriptor for the value of the corresponding identifier.

Consider the expression

```
j := 1
```

The corresponding virtual machine instructions are

```
pnull              # push null descriptor for the result
local     2        # push variable descriptor for j
int       1        # push descriptor for the integer 1
asgn               # perform assignment
```

When these instructions are interpreted, the succession of stack states is

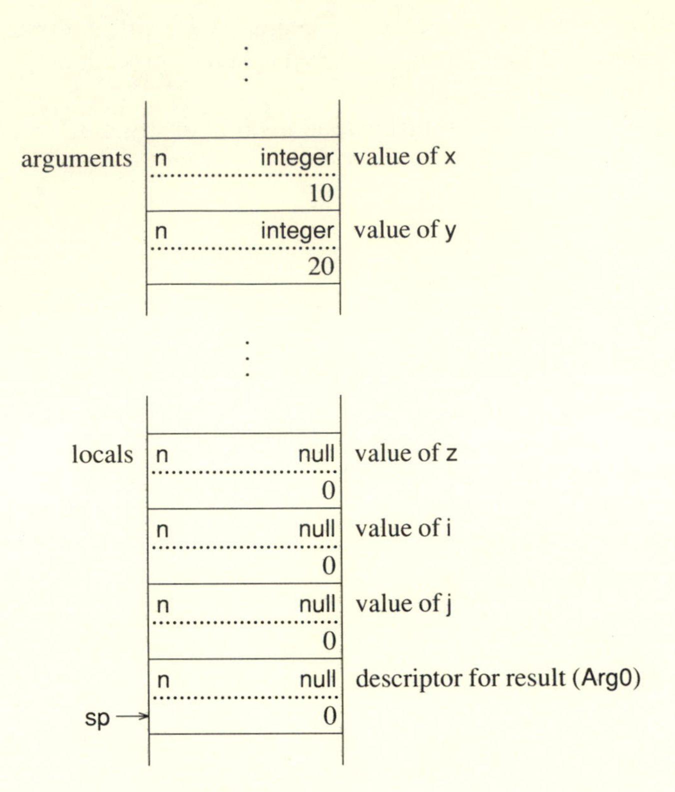


The Stack after pnull

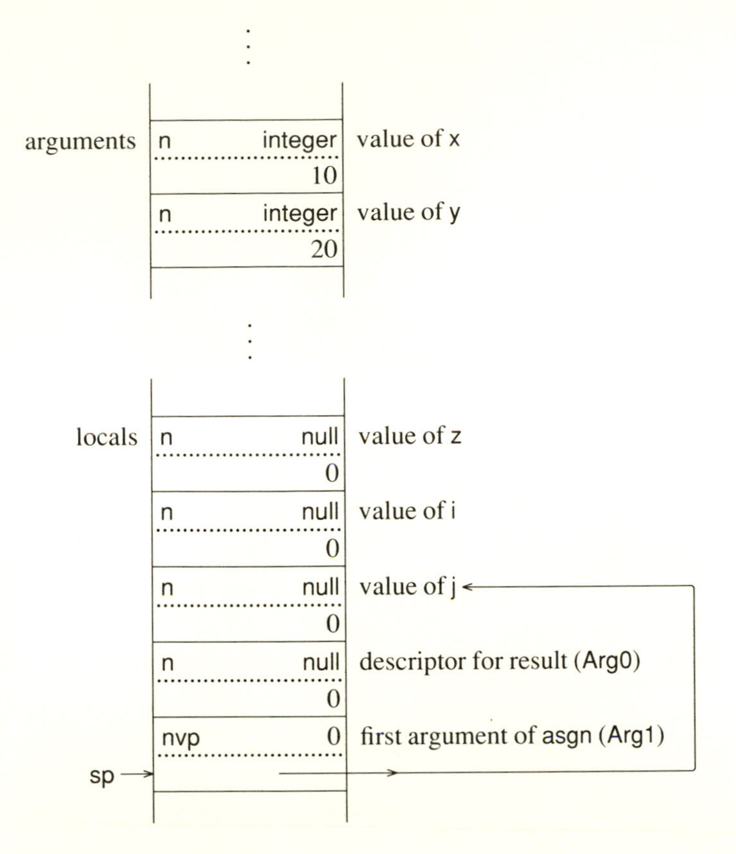


The Stack after local 2

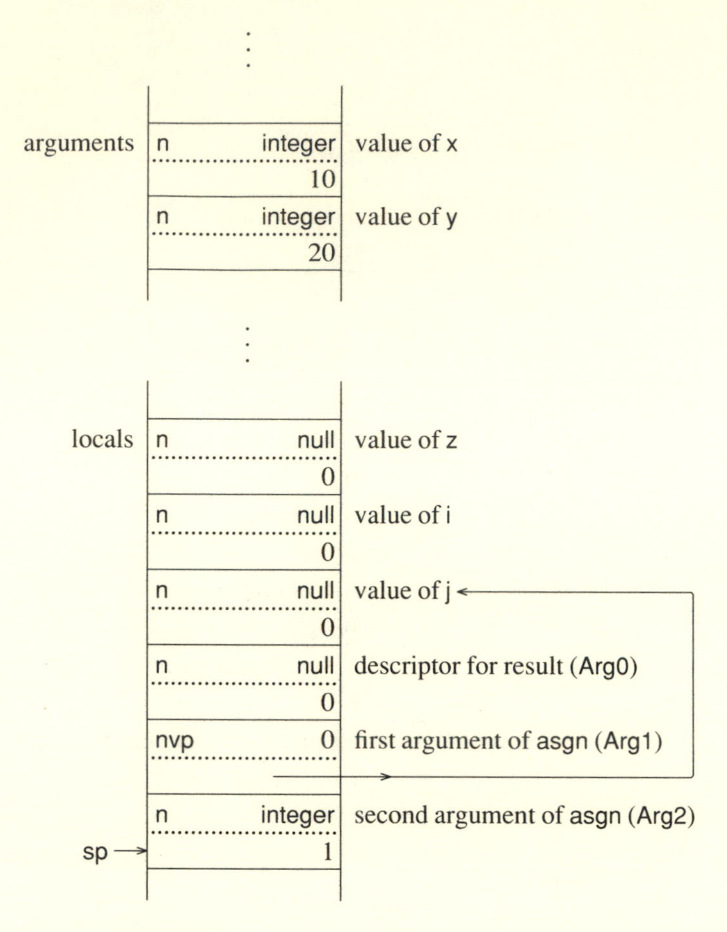


The Stack after int 1

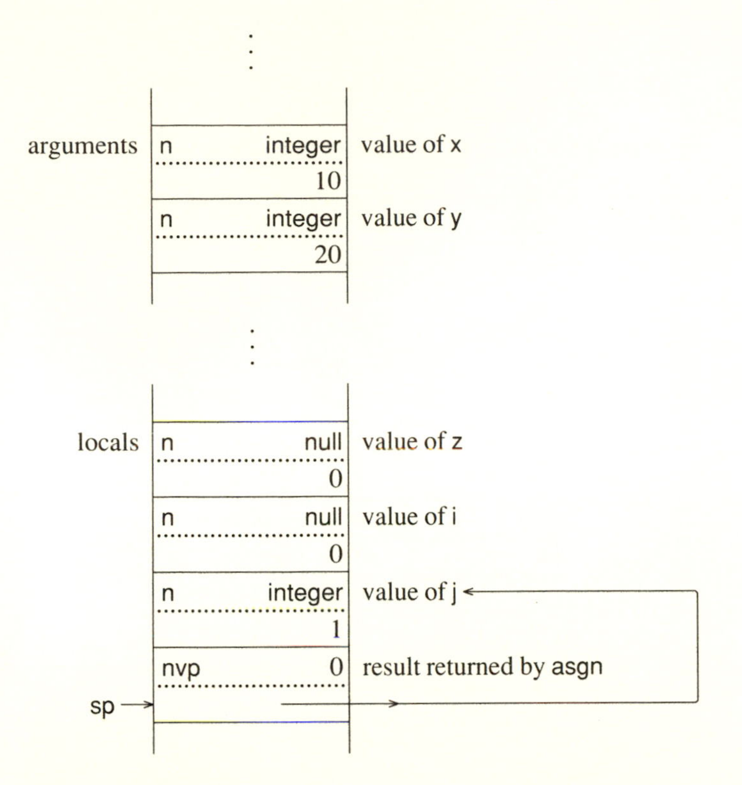


The Stack after asgn

Note that asgn assigns the value of its second argument to j and overwrites Arg0 with a variable descriptor, which is left on the top of the stack.

Similarly, the virtual machine instructions for

```
z := x
```

are

```
pnull
local    0
arg      0
asgn
```

and the states of the stack are

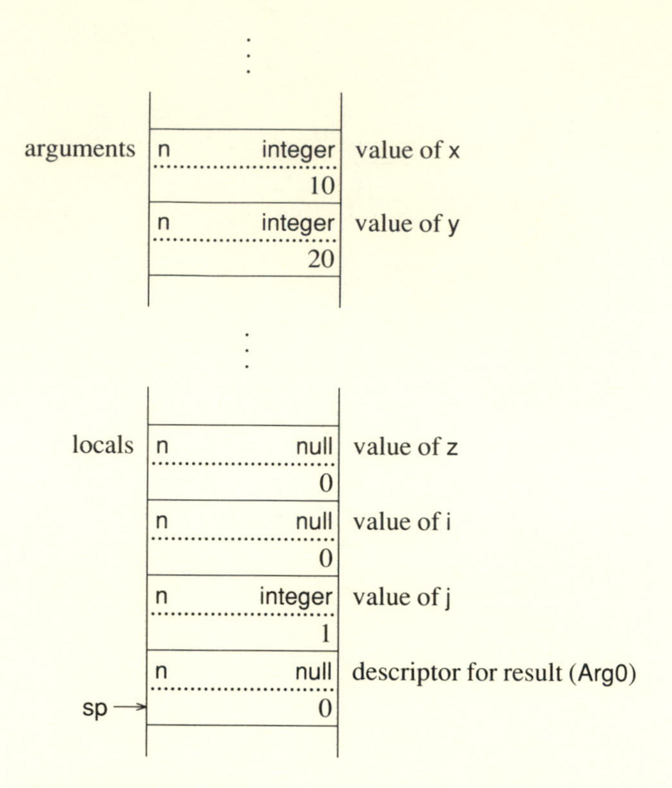


The Stack after pnull

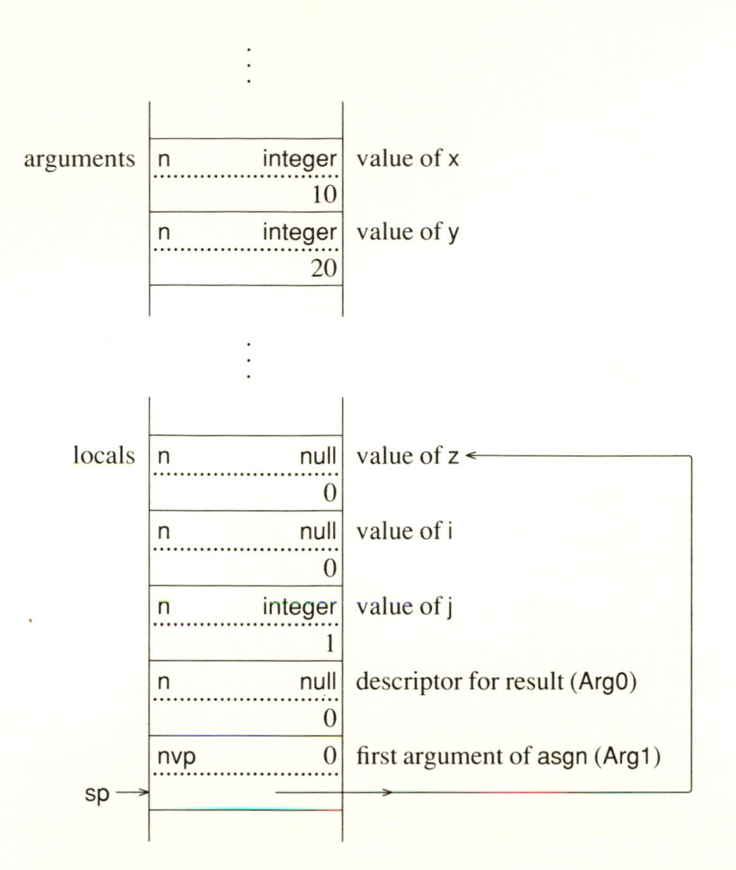


The Stack after local 0

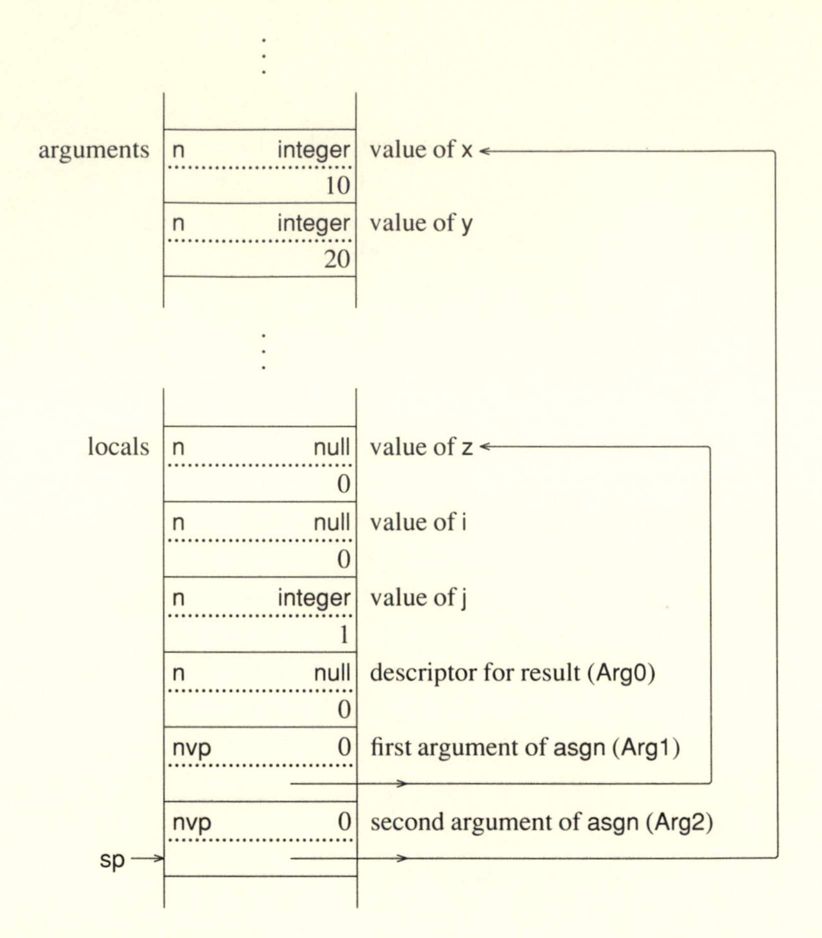


The Stack after arg 0

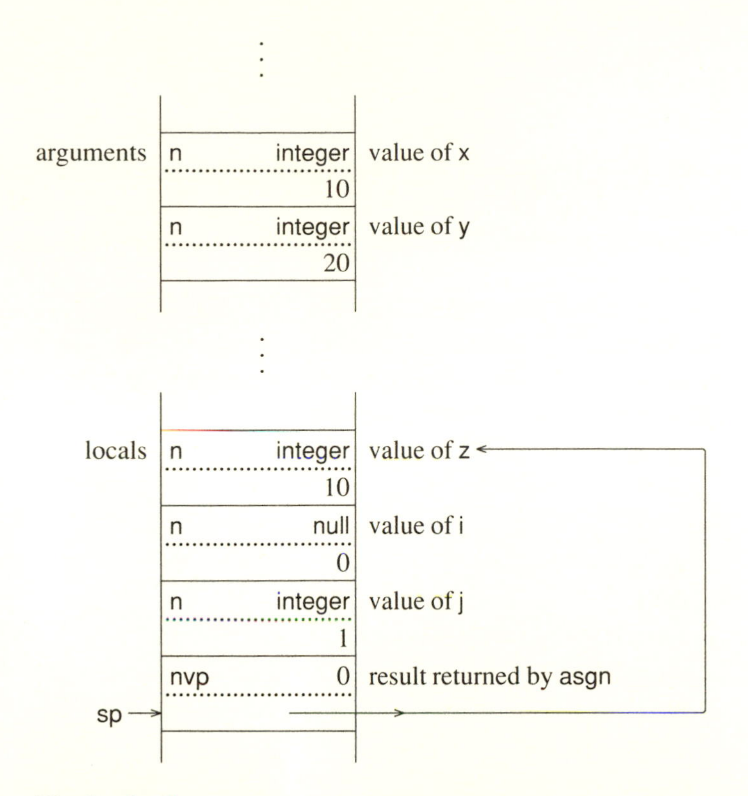


The Stack after asgn

### 8.2.3 Operators

There is a virtual machine instruction for each of the forty-two operators in Icon. The instructions random and asgn described previously are examples. Casting Icon operators as virtual machine instructions masks a considerable amount of complexity, since few Icon operators are simple. For example, although x + y appears to be a straightforward computation, it involves checking the types of x and y, converting them to numeric types if they are not already numeric, and terminating with an error message if this is not possible. If x and y are numeric or convertible to numeric, addition is performed. Even this is not simple, since the addition may be integer or floating-point, depending on the types of the arguments. For example, if x is an integer and y is a real number, the integer is converted to a real number. None of these computations is evident in the virtual machine instructions produced for this expression, which are

```
pnull
local    x
local    y
plus
```

In the instructions given previously, the indices that are used to access identifiers have been replaced by the names of the identifiers, which are assumed to be local. This convention is followed in subsequent virtual machine instructions for ease of reading.

Augmented assignment operations do not have separate virtual machine instructions. Instead, the instruction dup first pushes a null descriptor and then pushes a duplicate of the descriptor that was previously on top of the stack. For example, the virtual machine instructions for

```
i +:= 1
```

are

```
pnull
local    i
dup
int      1
plus
asgn
```

The stack after the execution of local is

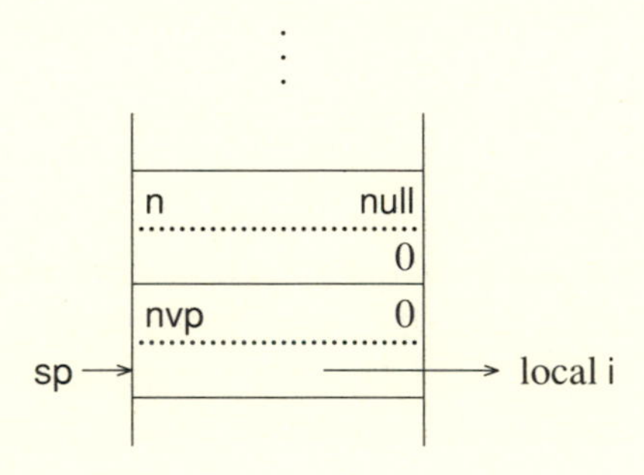


The execution of dup produces

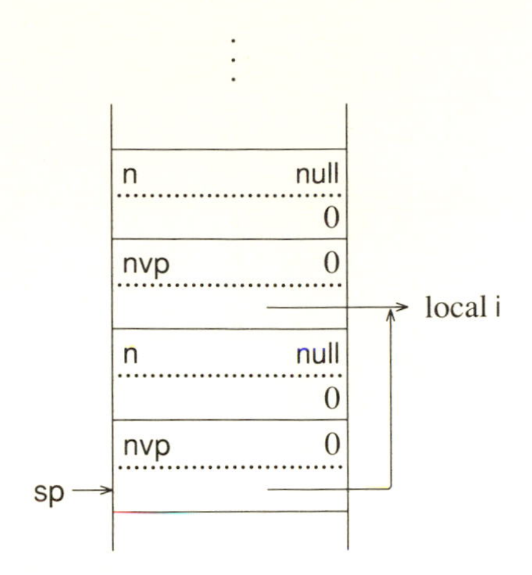


The dup instruction simply takes the place of the pnull and second local instructions in the virtual machine instructions for

```
i := i + 1
```

which are

```
pnull
local      i
pnull
local      i
int        1
plus
asgn
```

In this case, only a single local instruction is avoided. If the variable to which the assignment is made is not just an identifier but, instead, a more complicated construction, as in

```
a[j] +:= 1
```

substantial computation may be saved by duplicating the result of the first argument expression instead of recomputing it.

### 8.2.4 Functions

While the meaning of an operation is fixed and can be translated into a specific virtual machine instruction, the meaning of a function call can change during program execution. The value of the function also can be computed, as in

```
(p[i])(x, y)
```

The general form of a call is

$expr_0(expr_1, expr_2, \ldots, expr_n)$

The corresponding virtual machine instructions are

*code for* $expr_0$
*code for* $expr_1$
*code for* $expr_2$
⋮
*code for* $expr_n$
invoke    n

The invoke instruction is relatively complicated, since the value of $expr_0$ may be a procedure, an integer (for mutual evaluation), or even a value that is erroneous. Function invocation is discussed in detail in Chapter 10.

## 8.3 THE INTERPRETER PROPER

### 8.3.1 The Interpreter Loop

The interpreter, which is called interp, is basically simple in structure. It maintains a location in the icode (ipc) and begins by fetching the instruction pointed to by ipc and incrementing ipc to the next location. It then branches to a section of code for processing the virtual machine instruction that it fetched. The interpreter loop is

```
for (;;) {
   op = GetWord;
   switch (op) {
      .
      .
      .
      case Op_Asgn:
      .
      .
      .
      case Op_Plus:
      .
      .
      .
      }
   continue;
   .
   .
   .
}
```

where GetWord is a macro that is defined to be (*ipc++).

Macros are used extensively in the interpreter to avoid repetitious coding and to make the interpreter easier to read. The coding is illustrated by the case clause for the instruction plus:

```
case Op_Plus:                    /* e1 + e2 */
   Setup_Op(2);
   DerefArg(1);
   DerefArg(2);
   Call_Op;
   break;
```

Setup_Op(n) sets up a pointer to the address of Arg0 on the interpreter stack. The resulting code is

```
rargp = (struct descrip *)(sp - 1) - n;
```

The value of n is the number of arguments on the stack.

DerefArg(n) dereferences argument n. If it is a variable, it is replaced by its value. Thus, dereferencing is done in place by changing descriptors on the interpreter stack.

Call_Op calls the appropriate C function with a pointer to the interpreter stack as provided by Setup_Op(n). The function itself is obtained by looking up op in an array of pointers to functions. The code produced by Call_Op is

```
(*(optab[op]))(rargp);
sp = (word *)rargp + 1;
```

In the case where a C function produces a result, as plus always does, that result is placed in the Arg0 descriptor on the interpreter stack, as illustrated by the examples in Chapters 4 and 5. The interpreter adjusts sp to point to the v-word of Arg0. The break transfers control to the end of the switch statement, where a continue statement transfers control to the beginning of the interpreter loop, and the next instruction is fetched.

As illustrated earlier, some virtual machine instructions have operands, which follow the instructions in the icode. The interpreter code for such an instruction fetches its operands. An example is

```
int          n
```

The interpreter code for int is

```
case Op_Int:                    /* integer */
   PushVal(D_Integer);
   PushVal(GetWord);
   break;
```

PushVal(x) pushes x onto the interpreter stack. Thus, the descriptor for the integer is constructed by first pushing the constant D_Integer for the d-word and then pushing the fetched operand for the v-word.

### 8.3.2 Interpreter State Variables

The state of the interpreter is characterized by several variables, called *i-state variables*. Two i-state variables mentioned previously are sp, the interpreter stack pointer, and ipc, the interpreter "program counter."

The interpreter also pushes frames on the interpreter stack when procedures are called. Such frames are analogous to the frames pushed on the C stack when a C function is called, and contain information (typically i-state variables) that is saved when a procedure is called and restored when a procedure returns. There are other kinds of frames for special aspects of expression evaluation; these are described in Chapter 9. Pointers to frames are themselves i-state variables.

The proper maintenance of i-state variables is a central aspect of the interpreter and is discussed in detail in the next two chapters.

RETROSPECTIVE: The interpreter models, in software, the hardware of a cpu. The instruction fetch, the execution of operations, and the flow of control are basically the same as that in hardware execution, as is the use of a stack.

An interpreter offers great flexibility. It is easy to add virtual machine instructions, change existing ones, or change how they are implemented. Tracing and monitoring also are easy to add. The interpreter is machine-independent and portable. These advantages outweigh the loss in efficiency associated with emulating hardware in software.

## EXERCISES

**8.1** Why is it advantageous for the first argument of str to be the length of the string rather than its address?

**8.2** Show the states of the stack for the execution of the virtual machine instructions for the following Icon expressions:

```
i := i + 1
l := ???j
```

**8.3** Give an example for which

$expr_1$ := $expr_1$ + $expr_2$

produces a different result from

$expr_1$ +:= $expr_2$

**8.4** Describe, in general terms, what would be involved in adding a new operator to Icon.

**8.5** Describe, in general terms, what would be involved in adding a new kind of literal to Icon.

**8.6** Suppose that functions were bound at translation time instead of being source-language values. How might the virtual machine be modified to take advantage of such a feature? What effect would this have on the interpreter?

CHAPTER 9

# Expression Evaluation

PERSPECTIVE: The preceding chapter presents the essentials of the interpreter and expression evaluation as it might take place in a conventional programming language in which every expression produces exactly one result. For example, expressions such as

```
i := j
k := i + j
i +:= ?k
```

each produce a single result: they can neither fail nor can they produce sequences of results.

The one feature of Icon that distinguishes it most clearly from other programming languages is the capacity of its expression-evaluation mechanism to produce no result at all or to produce more than one result. From this capability come unconventional methods of controlling program flow, novel control structures, and goal-directed evaluation.

The generality of this expression-evaluation mechanism alone sets Icon apart from other programming languages. While generators, in one form or another, exist in a number of programming languages, such as IPL-V (Newell 1961), CLU (Liskov 1981), Alphard (Shaw 1981), and SETL (Dewar, Schonberg, and Schwartz 1981), such generators are limited to specific constructs, designated contexts, or restricted types of data. Languages with pattern-matching facilities, such as SNOBOL4 (Griswold, Poage, and Polonsky 1971), InterLisp (Teitelman 1974), and Prolog (Clocksin and Mellish 1981), generate alternative matches, but only within pattern matching.

Just as Icon's expression-evaluation mechanism distinguishes it from other programming languages, it is also one of the most interesting and challenging aspects of Icon's implementation. Its applicability in every context and to all kinds of data has a pervasive effect on the implementation.

## 9.1 BOUNDED EXPRESSIONS

A clear understanding of the semantics of expression evaluation in Icon is necessary to understand the implementation. One of the most important concepts of expression evaluation in Icon is that of a *bounded expression*, within which backtracking can take place. However, once a bounded expression has produced a result, it cannot be resumed for another result. For example, in

```
write(i = find(s1, s2))
```

find may produce a result and may be resumed to produce another result if the comparison fails. On the other hand, in

```
write(i = find(s1, s2))
write(j = find(s1, s3))
```

the two lines constitute separate expressions. Once the evaluation of the expression on the first line is complete, it cannot be resumed. Likewise, the evaluation of the expression on the second line is not affected by whether the expression on the first line succeeds or fails. However, if the two lines are joined by a conjunction operation, as in

```
write(i = find(s1, s2)) &
write(j = find(s1, s3))
```

they are combined into a larger single expression and the expression on the second line is not evaluated if the expression on the first line fails. Similarly, if the expression on the first line succeeds, but the expression on the second line fails, the expression on the first line is resumed.

The reason for the difference in the two cases is obscured by the fact that the Icon translator automatically inserts a semicolon at the end of a line on which an expression is complete and for which a new expression begins on the next line. Consequently, the first example is equivalent to

```
write(i = find(s1, s2));
write(j = find(s1, s3))
```

The difference between the semicolon and the conjunction operator is substantial. A semicolon bounds an expression, while an operator binds its operands into a single expression.

Bounded expressions are enclosed in ovals in the following examples to make the extent of backtracking clear. A compound expression, for example, has the following bounded expressions:

{ ($expr_1$;) ($expr_2$;) ...; $expr_n$ }

Note that $expr_n$ is not, of itself, a bounded expression. However, it may be part of a larger bounded expression, as in

{ $expr_1$; $expr_2$; ...; $expr_n$ } = 1;

Here $expr_n$ is part of the bounded expression for the comparison operator. The entire enclosing bounded expression is a consequence of the final semicolon. In the absence of the context provided by this semicolon, the entire expression might be part of a larger enclosing bounded expression, and so on.

The separation of a procedure body into a number of bounded expressions, separated by semicolons (explicit or implicit) and other syntactic constructions, is very important. Otherwise, a procedure body would consist of a single expression, and failure of any component would propagate throughout the entire procedure body. Instead, control backtracking is limited in scope to a bounded expression, as is the lifetime (and hence stack space) for temporary computations.

Bounded expressions are particularly important in control structures. For example, in the if-then-else control structure, the control expression is bounded but the other expressions are not:

if $expr_1$ then $expr_2$ else $expr_3$

As with the compound expression illustrated earlier, $expr_2$ or $expr_3$ (whichever is selected) may be the part of a larger bounded expression. An example is

```
write( if i < j then i to j else j to i )
```

If the control expression were not a separate bounded expression, the failure of $expr_2$ or $expr_3$ would result in backtracking into it and the if-then-else expression would be equivalent to

($expr_1$ & $expr_2$) | $expr_3$

which is hardly what is meant by if-then-else.

In a while-do loop, the control expression and the expression in the do clause are both bounded:

while $expr_1$ do $expr_2$

The two bounded expressions ensure that the expressions are evaluated independently of each other and any surrounding context. For example, if $expr_2$ fails, there is no control backtracking into $expr_1$.

### 9.1.1 Expression Frames

In the implementation of Icon, the scope of backtracking is delineated by *expression frames*. The virtual machine instruction

```
mark        L1
```

starts an expression frame. If the subsequent expression fails, ipc is set to the location in the icode that corresponds to L1. The value of ipc for a label is relative to the location of the icode that is read in from the icode file. For simplicity in the description that follows, the value of ipc is referred to just by the name of the corresponding label.

The mark instruction pushes an *expression frame marker* onto the stack and sets the expression frame pointer, efp, to it. Thus, efp indicates the beginning of the current expression frame. There is also a generator frame pointer, gfp, which points to another kind of frame that is used to retain information when an expression suspends with a result and is capable of being resumed for another. Generator frames are described in Sec. 9.3. The mark instruction sets gfp to zero, indicating that there is no suspended generator in a new expression frame.

An expression frame marker consists of four words: the value ipc for the argument of mark (called the failure ipc), the previous efp, the previous gfp, and ilevel, which is related to suspended generators:

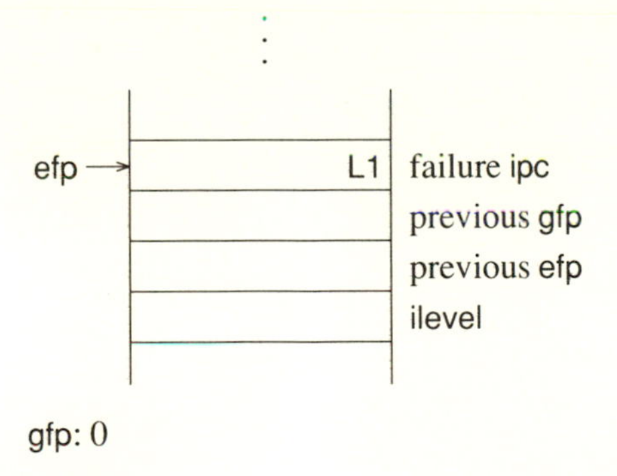


An expression frame marker is declared as a C structure:

```
struct ef_marker {              /* expression frame marker */
   word *ef_failure;            /*    failure ipc */
   struct ef_marker *ef_efp;    /*    efp */
   struct gf_marker *ef_gfp;    /*    gfp */
   word ef_ilevel;              /*    ilevel */
   };
```

This structure is overlaid on the interpreter stack in order to reference its components. The code for the mark instruction is

```
case Op_Mark:           /* create expression frame marker */
   newefp = (struct ef_marker *)(sp + 1);
   opnd = GetWord;
   opnd += (word)ipc;
   newefp->ef_failure = (word *)opnd;
   newefp->ef_gfp = gfp;
   newefp->ef_efp = efp;
   newefp->ef_ilevel = ilevel;
   sp += Wsizeof(*efp);
   efp = newefp;
   gfp = 0;
   break;
```

The macro Wsizeof(x) produces the size of x in words.

An expression frame is removed by the virtual machine instruction

```
unmark
```

which restores the previous efp and gfp from the current expression frame marker and removes the current expression frame by setting sp to the word just above the frame marker.

The use of mark and unmark is illustrated by

if ( $expr_1$ ) then $expr_2$ else $expr_3$

for which the virtual machine instructions are

```
      mark       L1
      code for expr1
      unmark
      code for expr2
      goto       L2
L1:
      code for expr3
L2:
```

The mark instruction creates an expression frame for the evaluation of $expr_1$. If $expr_1$ produces a result, the unmark instruction is evaluated, removing the expression frame for $expr_1$, along with the result produced by $expr_1$. Evaluation then proceeds in $expr_2$.

If *expr₁* fails, control is transferred to the location in the icode corresponding to L1 and the unmark instruction is not executed. In the absence of generators, failure also removes the current expression frame, as described in Sec. 9.2.

It is necessary to save the previous value of efp in a new expression marker, since expression frames may be nested. This occurs in interesting ways in some generative control structures, which are discussed in Sec. 9.4. Nested expression frames also occur as a result of evaluating compound expressions, such as

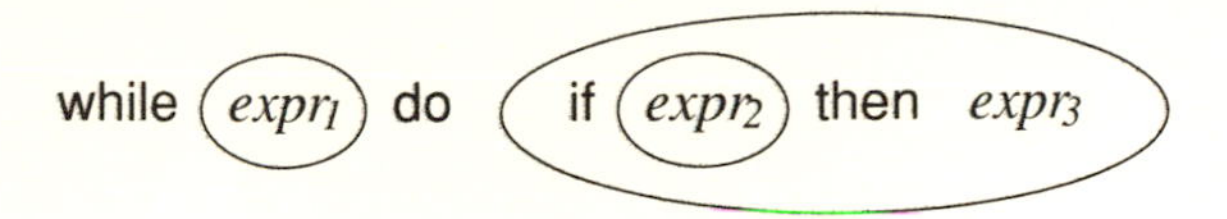


## 9.2 FAILURE

The interesting aspects of implementing expression evaluation in Icon can be divided into two cases: without generators and with generators. The possibility of failure in the absence of generators is itself of interest, since it occurs in other programming languages, such as SNOBOL4. This section describes the handling of failure and assumes, for the moment, that there are no generators. The next section describes generators.

In the absence of generators, if failure occurs anywhere in an expression, the entire expression fails without any further evaluation. For example, in the expressions

```
i := numeric(s)
line := read(f)
```

if numeric(s) fails in the first line, the assignment is not performed and evaluation continues immediately with the second line. In the implementation, this amounts to removing the current expression frame in which failure occurs and continuing with ipc set to the failure ipc from its expression frame marker.

The virtual machine instructions for the previous example are

```
          mark      L1
          pnull
          local     i
          global    numeric
          local     s
          invoke    1
          asgn
          unmark
     L1:
          mark      L2
          pnull
          local     line
          global    read
          local     f
          invoke    1
          asgn
          unmark
     L2:
```

Prior to the evaluation of the expression on the first line, there is some expression frame on the stack:

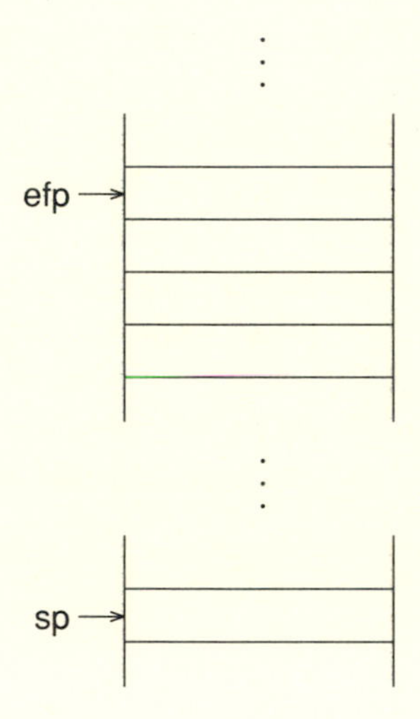


The instruction

```
          mark      L1
```

starts a new expression frame. The execution of subsequent virtual machine instructions pushes additional descriptors. The state of the stack when numeric is called by the invoke instruction is

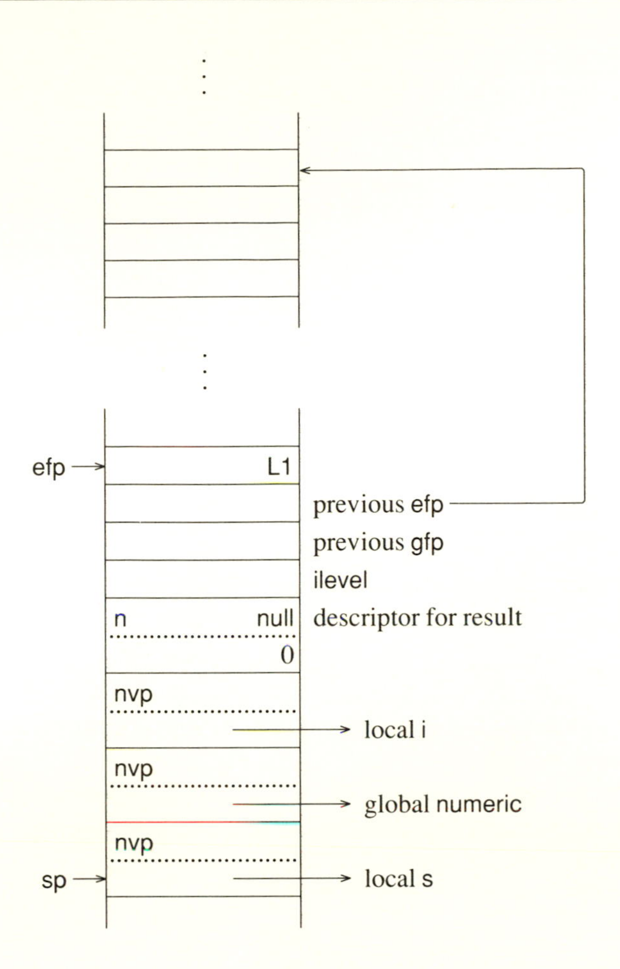


If numeric fails, efp and sp are reset, so that the stack is in the same state as it was prior to the evaluation of the expression on the first line:

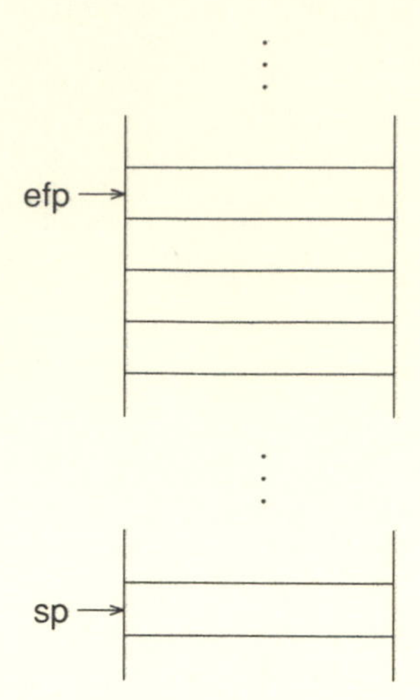


Control is transferred to the location in the icode corresponding to L1, and the execution of

```
mark        L2
```

starts a new expression frame by pushing a new expression frame marker onto the stack.

It is worth noting that failure causes only the current expression frame to be removed and changes ipc to the failure ipc. Any remaining virtual machine instructions in the current expression frame are bypassed; failure is simple and quick.

Failure can occur at three levels: directly from the virtual machine instruction efail, from a C function that implements an operator or function (as in the previous example), or from an Icon procedure.

When a conditional operator or function returns, it signals the interpreter, indicating whether it is producing a result or failing by using one of the two forms of return, Return or Fail. These macros simply produce return statements with different returned values.

The code in the interpreter for a conditional operation is illustrated by

```
case Op_Numlt:                    /* e1 < e2 */
   Setup_Op(2);
   DerefArg(1);
   DerefArg(2);
   Call_Cond;
   break;
```

The macro Call_Cond is similar to Call_Op described in Sec. 8.3.1, but it tests the signal returned by the C function. If the signal corresponds to the production of a result, the break is executed and control is transferred to the beginning of the interpreter loop to fetch the next virtual machine instruction. On the other hand, if the signal corresponds to failure, control is transferred to the place in the interpreter that handles failure, efail.

An Icon procedure can fail in three ways: by evaluating the expression fail, by the failure of the argument of a return expression, or by flowing off the end of the procedure body. The virtual machine instructions generated for the three cases are similar. For example, the virtual machine instructions for

```
if i < j then fail else write(j)
```

are

```
          mark      L1
          pnull
          local     i
          local     j
          numlt
          unmark
          pfail
L1:
          global    write
          local     j
          invoke    1
```

The virtual machine instruction pfail first returns from the current procedure call (see Sec. 10.3), and then transfers to efail.

## 9.3 GENERATORS AND GOAL-DIRECTED EVALUATION

The capability of an expression not to produce a result is useful for controlling program flow and for bypassing unneeded computation, but generators add the real power and expressiveness to the expression-evaluation semantics of Icon. It should be no surprise that generators also present difficult implementation problems. There are several kinds of generators, including those for control structures, functions and operators, and procedures. While the implementation of the different kinds of generators varies in detail, the same principles apply to all of them.

As far as using a result of an expression in further computation is concerned, there is no difference between an expression that simply produces a result and an expression that produces a result and is capable of being resumed to produce another one. For example, in

```
i := numeric("2")
j := upto('aeiou', "Hello world")
```

the two assignment operations are carried out in the same way, even though upto is a generator and numeric is not.

Since such contexts cannot be determined, in general, prior to the time the expressions are evaluated, the implementation is designed so that the interpreter stack is the same, as far as enclosing expressions are concerned, whether an expression returns or suspends. For the previous example, the arguments to the assignment operation are in the same relative place in both cases.

On the other hand, if a generator that has suspended is resumed, it must be capable of continuing its computation and possibly producing another result. For this to be possible, both the generator's state and the state of the interpreter stack must be preserved. For example, in

```
j := (i < upto('aeiou', "Hello world"))
```

when the function upto suspends, both i and the result produced by upto must be on the stack as arguments of the comparison operation. However, if the comparison operation fails and upto is resumed, the arguments of upto must be on the stack as they were when upto suspended. To satisfy these requirements, when upto suspends, a portion of the stack prior to the arguments for upto is copied to the top of the stack and the result produced by upto is placed on the top of the stack. Thus, the portion of the stack required for the resumption of upto is preserved and the arguments for the comparison are in the proper place.

**Generator Frames.** When an expression suspends, the state of the interpreter stack is preserved by creating a *generator frame* on the interpreter stack that contains a copy of the portion of the interpreter stack that is needed if the generator is resumed. A generator frame begins with generator frame marker that contains information about the interpreter state that must be restored if the corresponding generator is resumed. There are three kinds of generator frames that are distinguished by different codes:

| | |
|---|---|
| G_Csusp | suspension from a C function |
| G_Esusp | suspension from an alternation expression |
| G_Psusp | suspension from a procedure |

For the first two types of generators, the information saved in the generator frame marker includes the code for the type of the generator, the i-state variables efp, gfp, ipc, and the source-program line number at the time the generator frame is created:

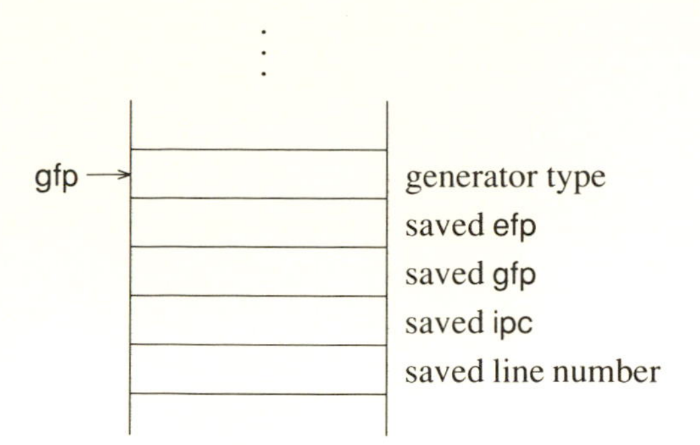


The corresponding C structure is

```
struct gf_marker {                    /* generator frame marker */
    word gf_gentype;                  /*   type */
    struct ef_marker *gf_efp;         /*   efp */
    struct gf_marker *gf_gfp;         /*   gfp */
    word *gf_ipc;                     /*   ipc */
    word gf_line;                     /*   line number */
};
```

Generators for procedure suspension contain, in addition, the i-state variables related to procedures. See Sec. 10.3.3.

As an example, consider the expression

```
write(i = (1 to 3));
```

The virtual machine instructions for this expression are

```
        mark      L1
        global    write
        pnull
        local     i
        int       1
        int       3
        push1               # default increment
        toby
        numeq
        invoke    1
        unmark
L1:
```

The state of the stack after execution of the first seven instructions is

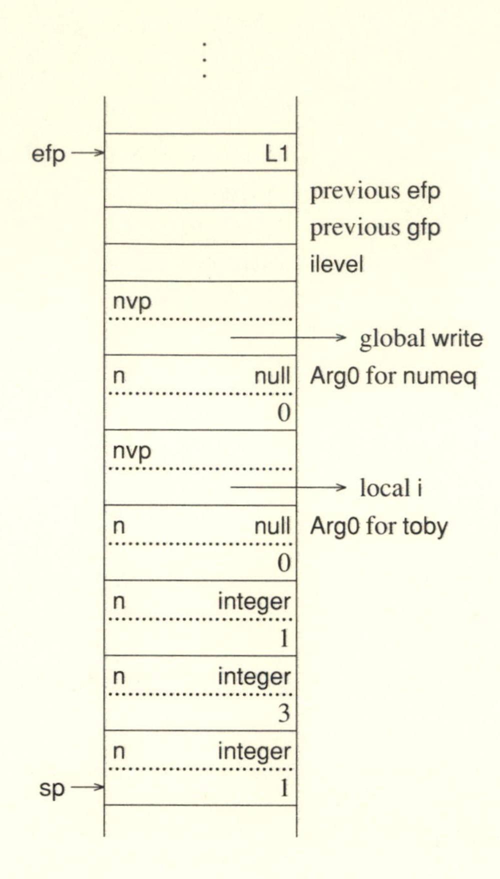


The code in the interpreter for calling a generative operator with n arguments is

```
rargp = (struct descrip *)(sp - 1) - n;
signal = (*(optab[op]))(rargp);
goto C_rtn_term;
```

Note that rargp points to Arg0 and is the argument of the call to the C function for the operator.

The C function for toby is

```
OpDcl(toby, 3, "toby")
   {
   long from, to, by;

   /*
    * Arg1 (from), Arg2 (to), and Arg3 (by) must be integers.
    *  Also, Arg3 must not be zero.
    */
   if (cvint(&Arg1, &from) == CvtFail)
      runerr(101, &Arg1);
   if (cvint(&Arg2, &to) == CvtFail)
      runerr(101, &Arg2);
   if (cvint(&Arg3, &by) == CvtFail)
      runerr(101, &Arg3);
   if (by == 0)
      runerr(211, &Arg3);

   /*
    * Count up or down (depending on relationship of from and to) and
    *  suspend each value in sequence, failing when the limit has been
    *  exceeded.
    */
   if (by > 0)
      for ( ; from <= to; from += by) {
         Mkint(from, &Arg0);    /* make an integer descriptor */
         Suspend;
         }
   else
      for ( ; from >= to; from += by) {
         Mkint(from, &Arg0);
         Suspend;
         }
   Fail;
   }
```

The OpDcl macro, which is similar to FncDcl, produces

```
toby(cargp)
register struct descrip *cargp;
```

so that toby is called with a pointer to Arg0. The macros Arg0, Arg1, and so forth are defined as

```
#define Arg0 (cargp[0])
#define Arg1 (cargo[1])
        :
        :
```

When toby is called, it replaces its Arg0 descriptor by a descriptor for the integer 1 and suspends by using the Suspend macro rather than Return.

The Suspend macro *calls* interp instead of returning to it. This leaves the call of toby intact with its variables preserved and also transfers control to interp, so that the next virtual machine instruction can be interpreted. However, it is necessary to push a generator marker on the interpreter stack and copy a portion of the interpreter stack, so that interpretation can continue without changing the portion of the interpreter stack that toby needs in case it is resumed. This is accomplished by calling interp with arguments that signal it to build a generator frame. The definition of Suspend is

```
#define Suspend { \
   int rc; \
   if ((rc = interp(G_Csusp, cargp)) != A_Resumption) \
      return rc; \
   }
```

The argument G_Csusp in the call of interp indicates that a generator frame for a C function is needed. The argument cargp points to the location on the interpreter stack where Arg0 for the suspending C function is located. This location is the same as rargp in the call of interp that called upto.

In this situation, interp puts a generator frame marker on the interpreter stack and copies the portion of the interpreter stack from the last expression or generator frame marker through cargp onto the top of the interpreter stack:

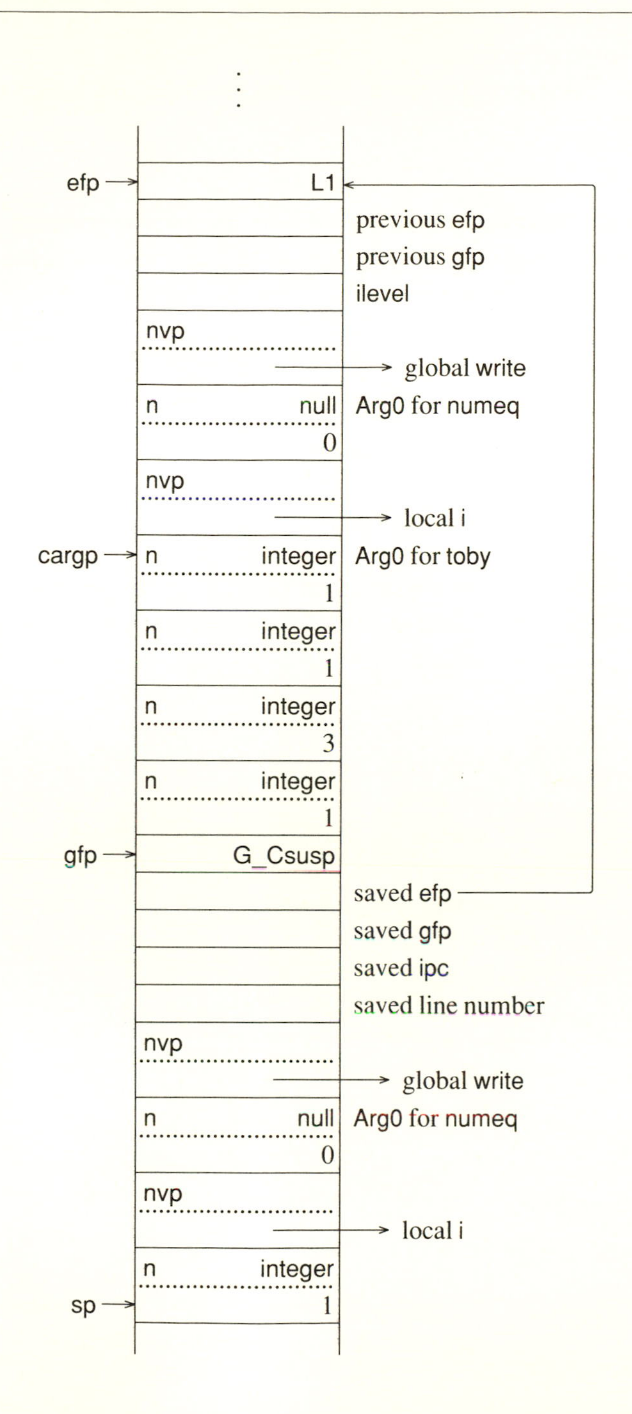


The stack is exactly the same, as far as the execution of numeq is concerned, as it would have been if toby had simply returned. However, the arguments of toby (and the preceding arguments of numeq) are still intact, so that toby can be resumed. The generator frame is interposed between the two portions of the interpreter stack. The top of the stack corresponds to the evaluation of

```
write(i = 1);
```

**Resumption.** Suppose the value of i in the previous example is 2. The comparison fails and control is transferred to efail, as it is in the case of all operations that fail. The code for efail is

```
                case Op_Efail:
efail:
                   /*
                    * Failure has occurred in the current expression frame.
                    */
                   if (gfp == 0) {
                      /*
                       * There are no suspended generators to resume.
                       *  Remove the current expression frame, restoring
                       *  values.
                       *
                       * If the failure ipc is 0, propagate failure to the
                       *  enclosing frame by branching back to efail.
                       *  This happens, for example, in looping control
                       *  structures that fail when complete.
                       */
                      ipc = efp->ef_failure;
                      gfp = efp->ef_gfp;
                      sp = (word *)efp - 1;
                      efp = efp->ef_efp;
                      if (ipc == 0)
                         goto efail;
                      break;
                      }

                   else {
                      /*
                       * There is a generator that can be resumed. Make
                       *  the stack adjustments and then switch on the
                       *  type of the generator frame marker.
                       */
                      register struct gf_marker *resgfp = gfp;

                      type = resgfp->gf_gentype;
                                    .
                                    .
                                    .
```

```
ipc = resgfp->gf_ipc;
efp = resgfp->gf_efp;
line = resgfp->gf_line;
gfp = resgfp->gf_gfp;
sp = (word *)resgfp - 1;
          .
          .
          .
switch (type) {

   case G_Csusp: {
      --ilevel;
      return A_Resumption;
      break;
      }

   case G_Esusp:
      goto efail;

   case G_Psusp:
      break;
   }

break;
}
```

If there were no generator frame (if gfp were 0), the entire expression frame would be removed, and the expression would fail as described in Sec. 9.2. However, since there is a C_Susp generator frame, the stack is restored to the state it was in when toby suspended, and the values saved in the generator frame marker are restored:

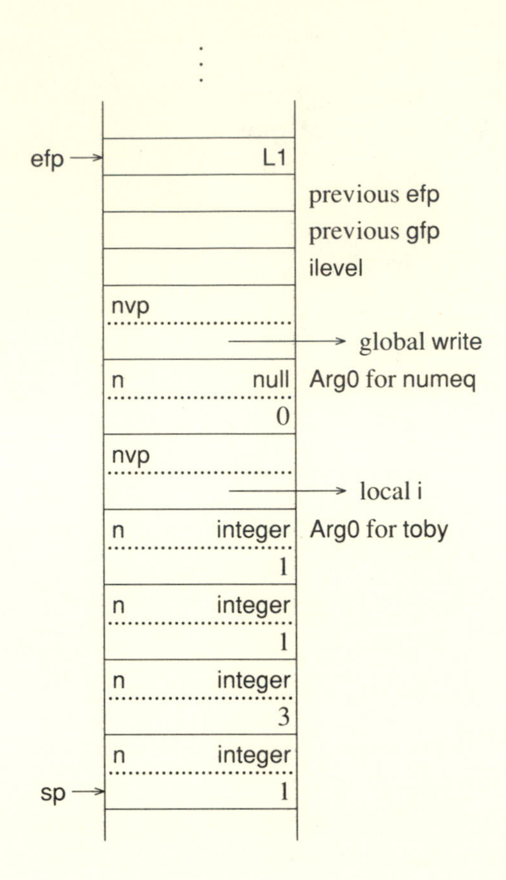


All traces of the first execution of numeq have been removed from the stack. As shown by the code for efail, the call to toby is resumed by *returning* to it from interp with the signal A_Resumption, which indicates another result is needed. When control is returned to toby, it changes its Arg0 descriptor to the integer 2 and suspends again:

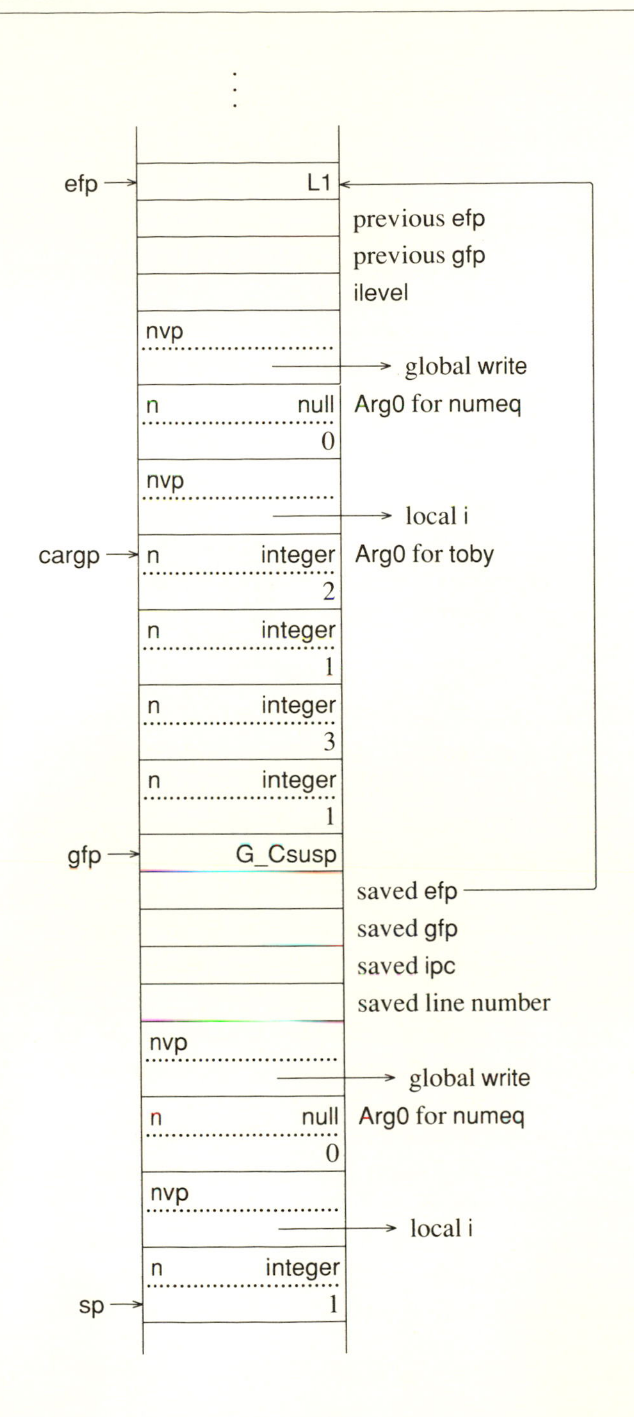


The interpreter stack is exactly as it was when toby suspended the first time, except that the integer 2 is on the stack in place of the integer 1. The top of the stack corresponds to the evaluation of

```
write(i = 2);
```

Since the value of i is 2, numeq succeeds. It copies the value of its second argument to its Arg0 descriptor and returns. The value 2 is written and the unmark instruction is executed, removing the entire expression frame from the stack.

**Goal-Directed Evaluation.** Goal-directed evaluation occurs when an expression fails and there are generator frames on the interpreter stack as the consequence of expressions that have suspended.

In the case of an expression such as

```
1 to upto(c, s)
```

upto suspends first, followed by toby. These generator frames are linked together, with gfp pointing to the one for toby, which in turn contains a pointer to the one for upto. In general, generator frames are linked together with gfp pointing to the one for the most recent suspension. This produces the last-in, first-out (depth-first) order of expression evaluation in Icon. Goal-directed evaluation occurs as a result of resuming a suspended expression when failure occurs in the surrounding expression frame.

**Removing C Frames.** Since C functions that suspend call the interpreter and the interpreter in turn calls C functions, expression evaluation typically results in a sequence of frames for calls on the C stack. When the evaluation of a bounded expression is complete, there may be frames on the C stack for generators, even though these generators no longer can be resumed.

In order to "unwind" the C stack in such cases, the i-state variable ilevel is used to keep track of the level of call of interp by C functions. Whenever interp is called, it increments ilevel. When an expression frame is created, the current value of ilevel is saved in it, as illustrated previously.

When the expression frame is about to be removed, if the current value of ilevel is greater than the value in the current expression frame, ilevel is decremented and the interpreter *returns* with a signal to the C function that called it to return rather than to produce another result. If the signal returned by interp is A_Resumption, the C function continues execution, while for any other signal the C function returns.

Since C functions return to interp, interp always checks the signal returned to it to determine if it produced a result or if it is unwinding. If it is unwinding, interp returns the unwinding signal instead of continuing evaluation of the current expression.

Consider again the expression

```
write(i = (1 to 3));
```

for which the virtual machine instructions are

```
        mark      L1
        global    write
        pnull
        local     i
        int       1
        int       3
        push1              # default increment
        toby
        numeq
        invoke    1
        unmark
L1:
```

When toby produces a result, it calls interp. When the unmark instruction is executed, the C stack contains a frame for the call to toby and for its call to interp. The code for unmark is

```
          case Op_Unmark:       /* remove expression frame */
             gfp = efp->ef_gfp;
             sp = (word *)efp - 1;

             /*
              * Remove any suspended C generators.
              */
Unmark_uw:
             if (efp->ef_ilevel < ilevel) {
                --ilevel;
                return A_Unmark_uw;
                }
             efp = efp->ef_efp;
             break;
```

Note that in this case Suspend gets the return code A_Unmark_uw and in turn returns A_Unmark_uw to interp. The section of code in interp that checks the signal that is returned from C functions is

```
C_rtn_term:
          switch (signal) {

             case A_Failure:
                goto efail;

             case A_Unmark_uw:          /* unwind for unmark */
                goto Unmark_uw;
```

```
            case A_Lsusp_uw:            /* unwind for lsusp */
               goto Lsusp_uw;

            case A_Eret_uw:             /* unwind for eret */
               goto Eret_uw;

            case A_Pret_uw:             /* unwind for pret */
               goto Pret_uw;

            case A_Pfail_uw:            /* unwind for pfail */
               goto Pfail_uw;
            }

         sp = (word *)rargp + 1;        /* set sp to result */
         continue;
         }
```

Thus, when interp returns to a C function with an unwinding signal, there is a cascade of C returns until ilevel is the same as it was when the current expression frame was created. Note that there are several cases in addition to unmark where unwinding is necessary.

## 9.4 GENERATIVE CONTROL STRUCTURES

In addition to functions and operators that may generate more than one result, there are several generative control structures at the level of virtual machine instructions.

### 9.4.1 Alternation

The virtual machine instructions for

$expr_2 \mid expr_3$

are

```
        mark      L1
        code for expr2
        esusp
        goto      L2
L1:
        code for expr3
L2:
```

The mark instruction creates an expression frame marker for alternation whose purpose is to preserve the failure ipc for L1 in case the results for *expr*$_3$ are needed. If *expr*$_2$ produces a result, esusp creates a generator frame with the usual marker and then copies the portion of the interpreter stack between the last expression or generator frame marker and the alternation marker to the top of the stack. It then pushes a copy of the result produced by *expr*$_2$. This connects the result produced by *expr*$_2$ with the expression prior to the alternation control structure. Next, esusp sets efp to point to the expression frame marker prior to the alternation marker. For example, in the expression

```
write(i := 1 | 2)
```

the stack after the execution of esusp is

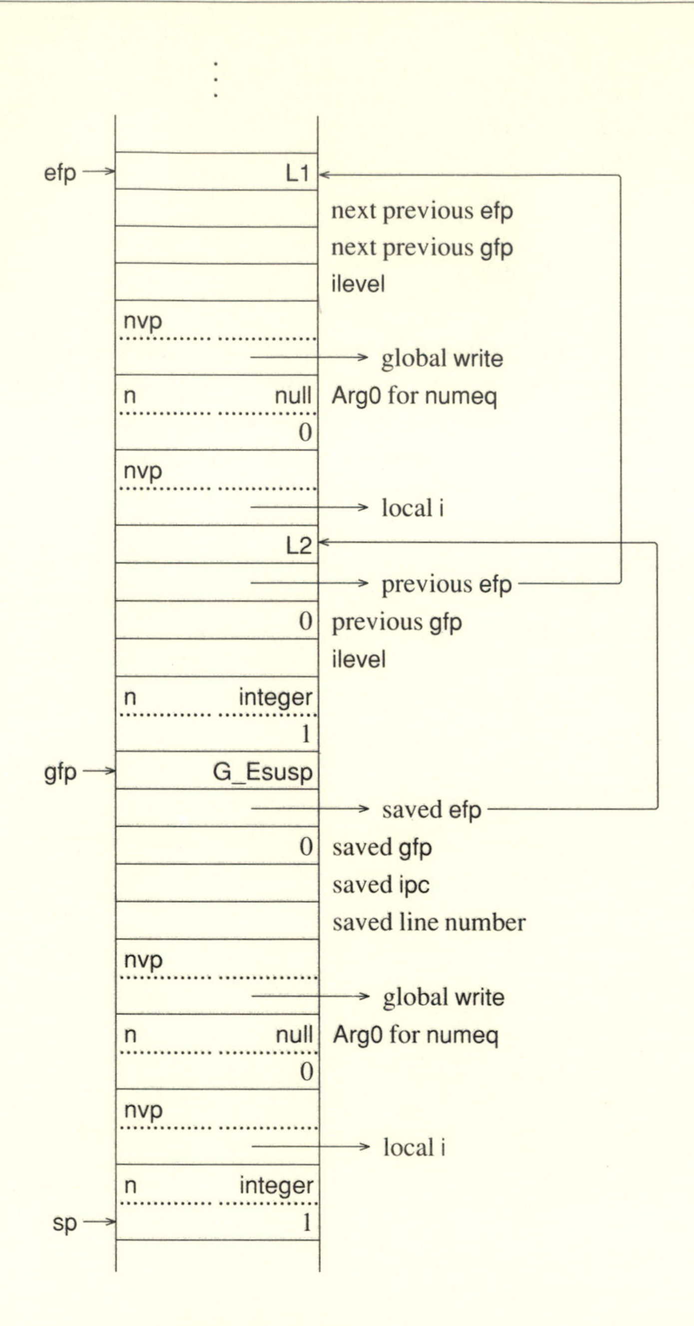


The top portion of the stack is the same as if $expr_2$ had produced a result in the absence of alternation. However, the generator frame marker pushed by esusp contains a pointer to the alternation marker.

If another result from $expr_2$ is needed, the generator frame left by esusp is removed, restoring the stack to its state when $expr_2$ produced a result. If $expr_2$ itself was a generator that suspended, it is resumed. Otherwise, control is transferred to efail and ipc is set to a value corresponding to L1, so that $expr_3$ is evaluated next.

### 9.4.2 Repeated Alternation

Alternation is the general model for generative control structures. Repeated alternation, |*expr*, is similar to alternation, and would be equivalent to

*expr* | *expr* | *expr* | ...

except for a special termination condition that causes repeated alternation to stop if *expr* produces a result. Without this termination condition, an expression such as

```
|upto(c, s)
```

would never return if upto failed—expression evaluation would vanish into a "black hole." Expressions that produce results at one time but not at another also are useful. For example,

```
|read()
```

generates the lines from the standard input file. Because of the termination condition, this expression terminates when the end of the input file is reached. If it vanished into a "black hole," it could not be used safely.

If it were not for the termination condition, the virtual machine instructions for |*expr* would be

```
L1:
        mark        L1
        code for expr
        esusp
```

The "black hole" here is evident—if *expr* fails, it is evaluated again and there is no way out.

The termination condition is handled by an instruction that changes the failure ipc in the current expression marker. The actual virtual machine instructions for |*expr* are

```
L1:
        mark0
        code for expr
        chfail      L1
        esusp
```

The virtual machine instruction mark0 pushes an expression frame marker with a zero failure ipc. If a zero failure ipc is encountered during failure, as illustrated by the code for efail in Sec. 9.3, failure is transmitted to the enclosing expression. If *expr* produces a result, however, the chfail instruction is executed. It changes the failure ipc in the current expression marker to correspond to L1, so that if *expr* does not produce a result when it is resumed, execution starts at the location in the icode corresponding to L1 again, causing another iteration of the alternation loop. It is important to realize that chfail only changes the failure ipc in the current expression marker on the stack. Subsequent execution of mark0 creates a new expression frame whose marker has a zero failure ipc.

### 9.4.3 Limitation

In the limitation control structure,

$expr_1 \setminus expr_2$

the normal left-to-right evaluation of Icon is reversed and $expr_2$ is evaluated first. The virtual machine instructions are

```
code for expr2
limit
code for expr1
lsusp
```

If $expr_2$ succeeds, its result is on the top of the stack. The limit instruction checks this result to be sure that it is legal—an integer greater than or equal to zero. If it is not an integer, an attempt is made to convert it to one. If the limit value is zero, limit fails. Otherwise, limit creates an expression frame marker with a zero failure ipc and execution continues, so that $expr_1$ is evaluated in its own expression frame. During the evaluation of $expr_1$, the limit value is directly below its expression marker. For example, in

$expr_1 \setminus 10$

the stack prior to the evaluation of $expr_1$ is

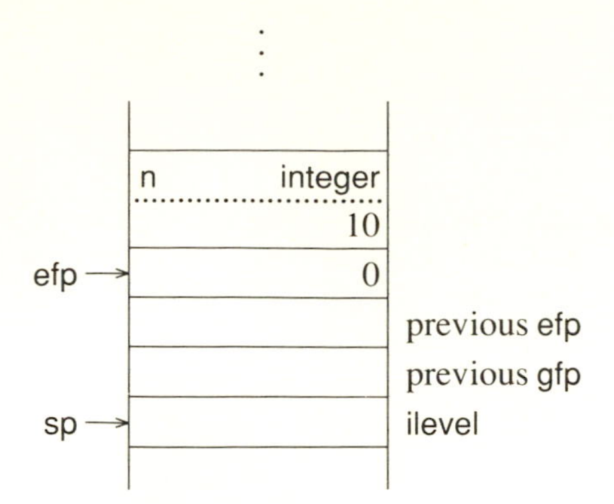


If $expr_1$ produces a result, lsusp is executed. The lsusp instruction is very similar to esusp. Before producing a generator frame, however, lsusp decrements the limit value. If it becomes zero, the expression frame for $expr_1$ is removed, the C stack is unwound, and the last value it produced is placed on the stack in place of the limit value. Otherwise, it copies the portion of the interpreter stack between the end of the last expression or generator frame marker and the limit value to the top of the stack. Note that no generator frame is needed.

## 9.5 ITERATION

The difference between evaluation and resumption in a loop is illustrated by the virtual machine instructions for a conventional loop

while $expr_1$ do $expr_2$

and the iteration control structure

every $expr_1$ do $expr_2$

The instructions for while-do are

```
L1:
        mark0
        code for expr1
        unmark
        mark      L1
        code for expr2
        unmark
        goto      L1
```

If $expr_1$ fails, the entire expression fails and failure is transmitted to the enclosing expression frame because the failure ipc is zero. If $expr_1$ produces a result, $expr_2$ is evaluated in a separate expression frame. Whether $expr_2$ produces a result or not, its expression frame is removed and execution continues at the beginning of the loop.

The instructions for every-do are

```
mark0
code for expr1
pop
mark0
code for expr2
unmark
efail
```

If $expr_1$ fails, the entire expression fails and failure is transmitted to the enclosing expression frame as in the case of while-do. If $expr_1$ produces a result, it is discarded by the pop instruction, since this result is not used in any subsequent computation. The expression frame for $expr_1$ is not removed, however, and $expr_2$ is evaluated in its own expression frame within the expression frame for $expr_1$ (unlike the case for the while loop). If $expr_2$ produces a result, its expression frame is removed and efail is executed. If $expr_2$ fails, it transmits failure to the enclosing expression frame, which is the expression frame for $expr_1$. If $expr_2$ produces a result, efail causes failure in the expression frame for $expr_1$. Thus, the effect is the same, whether or not $expr_2$ produces a result. All results are produced simply by forcing failure.

If the expression frame for $expr_1$ contains a generator frame, which is the case if $expr_1$ suspended, evaluation is resumed accordingly, so that $expr_1$ can produce another result. If $expr_1$ simply produces a result instead of suspending, there is no generator frame, efail removes its expression frame, and failure is transmitted to the enclosing expression frame.

## 9.6 STRING SCANNING

String scanning is one of the most useful operations in Icon. Its implementation, however, is comparatively simple. There is no special expression-evaluation mechanism associated with string scanning *per se*; all ''pattern matching'' follows naturally from goal-directed evaluation.

The string-scanning keywords, &subject and &pos must be handled properly, however. These keywords have global scope with respect to procedure invocation, but they are maintained in a stack-like fashion with respect to string-scanning expressions.

The expression

$expr_1$ ? $expr_2$

is a control structure, not an operation, since, by definition, the arguments for an operation are evaluated before the operation is performed. This form of evaluation does not work for string scanning, since after $expr_1$ is evaluated, but before $expr_2$ is evaluated, the previous values of &subject and &pos must be saved and new ones established. Furthermore, when string scanning is finished, the old values of &subject and &pos must be restored. In addition, if string scanning succeeds, the values of these keywords at the time string scanning produces a result must be saved so that they can be restored if the string-scanning operation is resumed to produce another result.

The virtual machine instructions for

$expr_1$ ? $expr_2$

are

*code for* $expr_1$
bscan
*code for* $expr_2$
escan

If $expr_1$ succeeds, it leaves a result on the top of the stack. The bscan instruction assures that this value is a string, performing a conversion if necessary. Otherwise, the old values of &subject and &pos are pushed on the stack, the value of &subject is set to the (possibly converted) one produced by $expr_1$, and &pos is set to 1.

The bscan instruction then suspends. This is necessary in case $expr_2$ fails, so that bscan can get control again to perform data backtracking, restoring the previous values of &subject and &pos from the stack where they were saved.

If $expr_2$ succeeds, the escan instruction copies the descriptor on the top of the stack to its Arg0 position, overwriting the result produced by $expr_2$. It then exchanges the current values of &subject and &pos with those saved by bscan, thus restoring the values of these keywords to their values prior to the scanning expression and at the same time saving the values they had at the time $expr_2$ produced a result. The escan instruction then suspends.

If escan is resumed, the values of &subject and &pos are restored from the stack, restoring the situation to what it was when $expr_2$ produced a result. The escan instruction then fails in order to force the resumption of any suspended generators left by $expr_2$.

Suppose, for example, that the values of &subject and &pos are "the" and 2, respectively, when the following expression is evaluated:

```
read(f) ? move(4)
```

Suppose read(f) produces the string "coconuts". The stack is

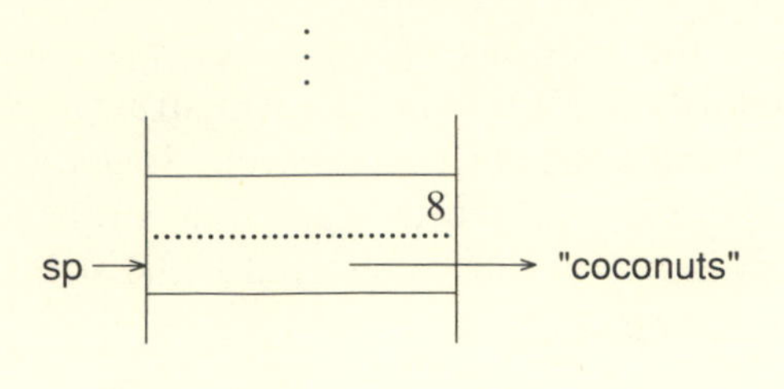


&subject: "the"
&pos: 2

The bscan instruction is executed, pushing the current values of &subject and &pos:

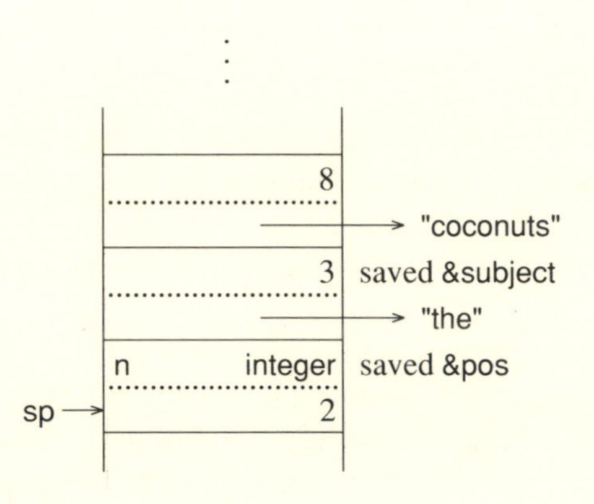


&subject: "the"
&pos: 2

The bscan instruction sets &subject to "coconuts" and &pos to 1. The bscan instruction then suspends and move(4) is evaluated. It suspends, and the top of the stack is

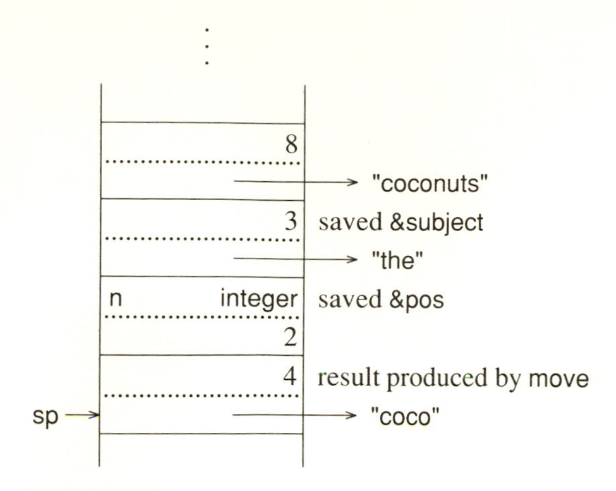


&subject: "coconuts"
&pos: 5

The escan instruction is executed next. It copies the descriptor on the top of the stack to replace the result produced by $expr_2$. It then exchanges the current values of &subject and &pos with those on the stack:

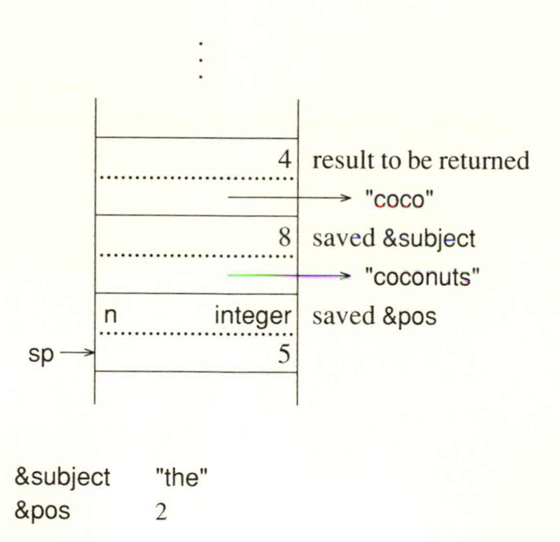


The escan instruction then suspends, building a generator frame. The result of $expr_3$ is placed on the top of the stack, becoming the result of the entire scanning expression:

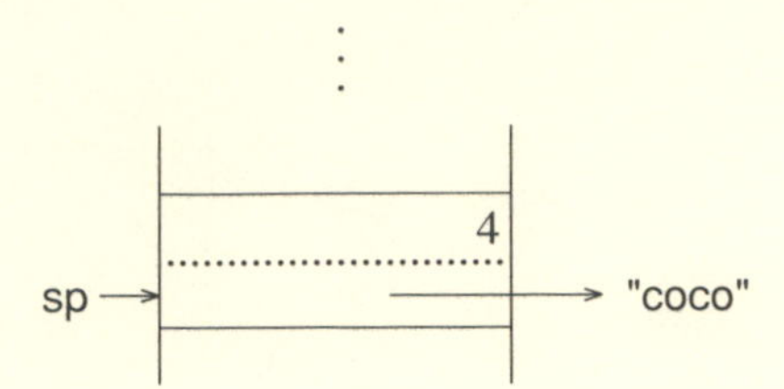


Since escan suspends, the saved values of &subject and &pos are preserved in a generator frame on the stack until escan is resumed or until the current expression frame is removed.

RETROSPECTIVE: The implementation of expression evaluation in Icon focuses on the concept of an expression frame within which control backtracking can occur. Bounded expressions, for example, are represented on the stack by expression frames, which confine backtracking.

In the absence of generators, failure simply results in the removal of the current expression frame and transfer to a new location in the icode, bypassing instructions that otherwise would have been executed.

State information must be saved when a generator suspends, so that its evaluation can be resumed. This information is preserved in a generator frame within the current expression frame. Generator frames are linked together in a last-in, first-out fashion. Goal-directed evaluation is a natural consequence of resuming the most recently suspended generator when an expression fails, instead of simply removing the current expression frame.

String scanning involves saving and restoring the values of &subject and &pos. This is somewhat complicated, since scanning expressions can suspend and be resumed, but string scanning itself introduces nothing new into expression evaluation: generators and goal-directed evaluation provide ''pattern matching.''

## EXERCISES

**9.1** Circle all the bounded expressions in the following segments of code:

```
while line := read() do
   if *line = i then write(line)

if (i = find(s1, s2)) & (j = find(s1, s3)) then {
   write(i)
   write(j)
   }
```

```
line ? while write(move(1)) do
   move(1)
```

**9.2** Describe the effect of nested generators and generators in mutual evaluation on the interpreter level.

**9.3** Consider a hypothetical control structure called *exclusive alternation* that is the same as regular alternation, except that if the first argument produces at least one result, the results from the second argument are not produced. Show the virtual machine instructions that should be generated for exclusive alternation.

**9.4** The expression read(f) is an example of an expression that may produce a result at one time and fail at another. This is possible because of a side effect of evaluating it—changing the position in the file f. Give an example of an expression that may fail at one time and produce a result at a subsequent time.

**9.5** There are potential ''black holes'' in the expression-evaluation mechanism of Icon, despite the termination condition for repeated alternation. Give an example of one.

**9.6** The expression frame marker produced by limit makes it easy to locate the limitation counter. Show how the counter could be located without this marker.

**9.7** Suppose that the virtual machine instructions for

every $expr_1$ do $expr_2$

did not pop the result produced by $expr_1$. What effect would this have?

**9.8** The virtual machine instructions for

every *expr*

are

```
mark     0
code for expr
pop
efail
```

so that failure causes *expr* to be resumed. The keyword &fail also fails, so that the virtual machine instructions for

*expr* & &fail

are

*code for expr*
efail

It is sometimes claimed that these two expressions are equivalent. If this were so, the shorter virtual machine instruction sequence for the second expression could be used for the first expression. Explain why the two expressions are not equivalent, in general, and give an example in which they are different.

**9.9** Diagram the states of the stack for the example given in Sec. 9.6, showing all generator frames.

**9.10** Show the successive stack states for the evaluation of the following expressions, assuming that the values of &subject and &pos are "the" and 2, respectively, and that read() produces "coconuts" in each case:

(a) read(f) ? move(10)
(b) (read(f) ? move(4)) ? move(2)
(c) read(f) ? (move(4) ? move(2))
(d) (read(f) ? move(4)) ? move(10)
(e) (read(f) ? move(4 | 6)) ? move(5)
(f) (read(f) ? move(4)) & (read(f) & move(10))

**9.11** Write Icon procedures to emulate string scanning. *Hint:* consider the virtual machine instructions for

$expr_1$ ? $expr_2$

CHAPTER 10

# Functions, Procedures, and Co-Expressions

PERSPECTIVE: The invocation of functions and procedures is central to the evaluation of expressions in most programming languages. In Icon, this activity has several aspects that complicate its implementation. Functions and procedures are data values that can be assigned to identifiers, passed as arguments, and so on. Consequently, the meaning of an invocation expression cannot be determined until it is evaluated. Functions and procedures can be called with more or fewer arguments than expected. Thus, there must be a provision for adjusting argument lists at run time. Since mutual evaluation has the same syntax as function and procedure invocation, run-time processing of such expressions is further complicated.

Co-expressions, which require separate stacks for their evaluation, add complexities and dependencies on computer architecture that are not found elsewhere in the implementation.

## 10.1 INVOCATION EXPRESSIONS

As mentioned in Sec. 8.2.4, the virtual machine code for an expression such as

$expr_0(expr_1, expr_2, \ldots, expr_n)$

is

*code for* $expr_0$
*code for* $expr_1$
*code for* $expr_2$
⋮
*code for* $expr_n$
invoke n

Consequently, the stack when the invoke instruction is executed is

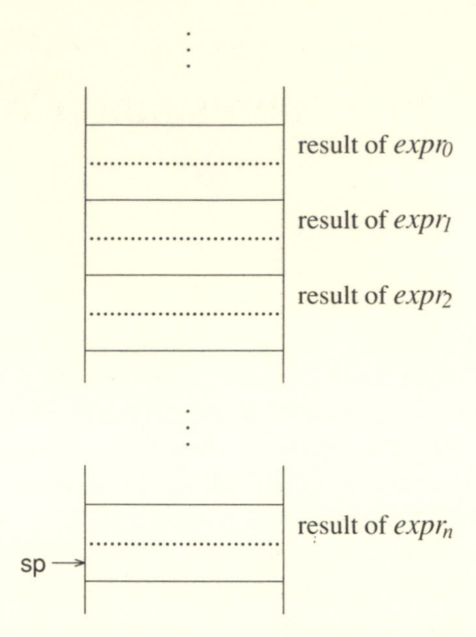


The meaning of the expression, and hence the action taken by invoke, depends on the result produced by $expr_0$. If the value of $expr_0$ is an integer or convertible to an integer, the invocation expression corresponds to mutual evaluation. If this integer is negative, it is converted to the corresponding positive value with respect to the number of arguments. If the value is between one and n, the corresponding descriptor is copied on top of the result of $expr_0$, sp is set to this position, and invoke transfers control to the beginning of the interpretive loop. On the other hand, if the value is out of range, invoke fails. Note that the returned value overwrites the descriptor for $expr_0$, whereas for operators a null-valued descriptor is pushed to receive the value.

If the value of $expr_0$ is a function or a procedure, the corresponding function or procedure must be invoked with the appropriate arguments. A function or procedure value is represented by a descriptor that points to a block that contains information about the function or procedure.

## 10.2 PROCEDURE BLOCKS

Functions and procedures have similar blocks, and there is no source-language type distinction between them.

**Blocks for Procedures.** Blocks for procedures are constructed by the linker, using information provided by the translator. Such blocks are read in as part of the icode file when an Icon program is executed. The block for a procedure contains the usual title and size words, followed by six words that characterize the procedure:

(1) The icode location of the first virtual machine instruction for the procedure.
(2) The number of arguments expected by the procedure.
(3) The number of local identifiers in the procedure.
(4) The number of static identifiers in the procedure.
(5) The index in the static identifier array of the first static identifier in the procedure.
(6) A C string for the name of the file in which the procedure declaration occurred.

The remainder of the procedure block contains qualifiers: one for the string name of the procedure, then others for the string names of the arguments, local identifiers, and static identifiers, in that order.

For example, the procedure declaration

```
procedure calc(i, j)
   local k
   static base, index
         :
         :
end
```

has the following procedure block:

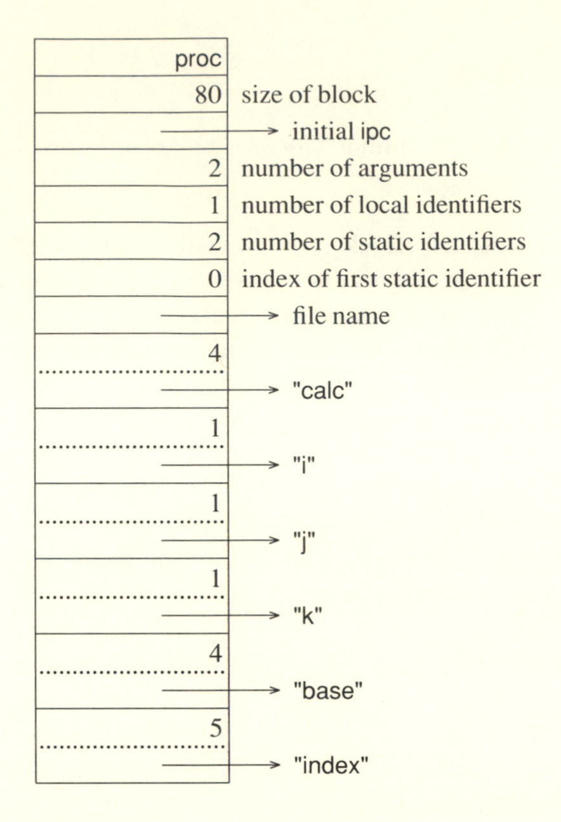


The 0 value for the index in the static identifier array indicates that base is the first static identifier in the program. The indices of static identifiers are zero-based and increase throughout a program as static declarations occur.

**Blocks for Functions.** Blocks for functions are created by the macro FncDcl that occurs at the beginning of every C function that implements an Icon function. Such blocks for functions are similar to those for procedures but are distinguished by the value −1 in the word that otherwise contains the number of local identifiers. The entry point is the entry point of the C routine for the function. The procedure block for repl is typical:

| | |
|---|---|
| proc | |
| 40 | size of block |
| → | C entry point |
| 2 | number of arguments |
| −1 | function indicator |
| 0 | not used |
| 0 | not used |
| 0 | not used |
| 4 | |
| → | "repl" |

Note that there are no argument names.

Some functions, such as write, allow an arbitrary number of arguments. This is indicated by the value −1 in place of the number of arguments:

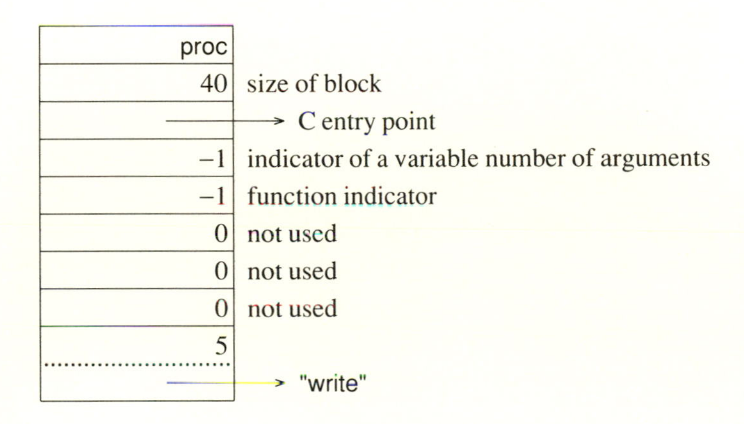


## 10.3 INVOCATION

### 10.3.1 Argument Processing

Argument processing begins by dereferencing the arguments in place on the stack. If a fixed number of arguments is specified in the procedure block, this number is compared with the argument of invoke, which is the number of arguments on the stack.

If there are too many arguments, sp is set to point to the last one expected. For example, the expression

numeric(i, j)

results in

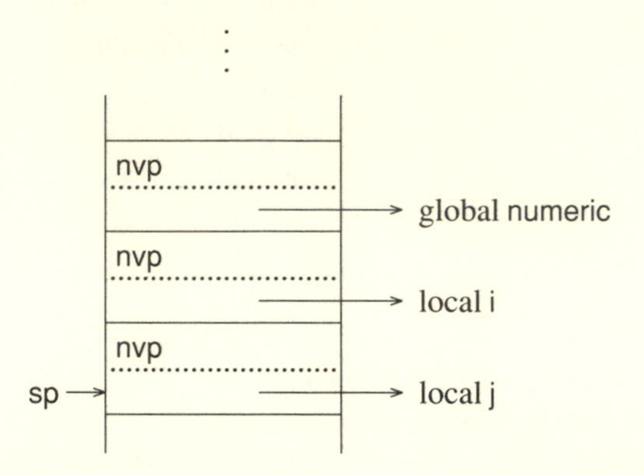


Since numeric expects only one argument, sp is reset, effectively popping the second argument:

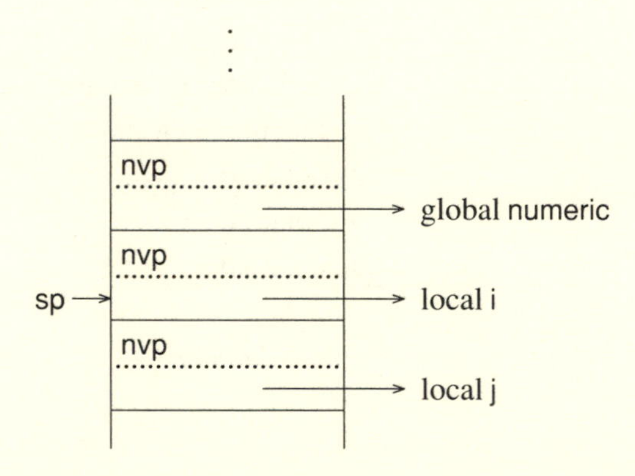


On the other hand, if there are not enough arguments, null-valued descriptors are pushed to supply the missing arguments. For example, the expression

left(s, i)

results in

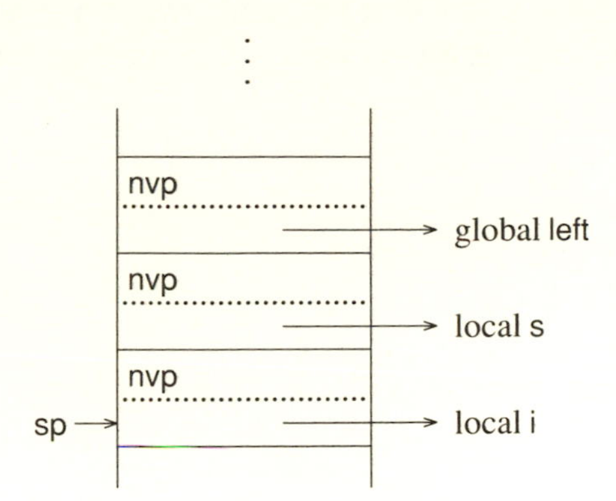


and a null value is pushed to provide the missing third argument:

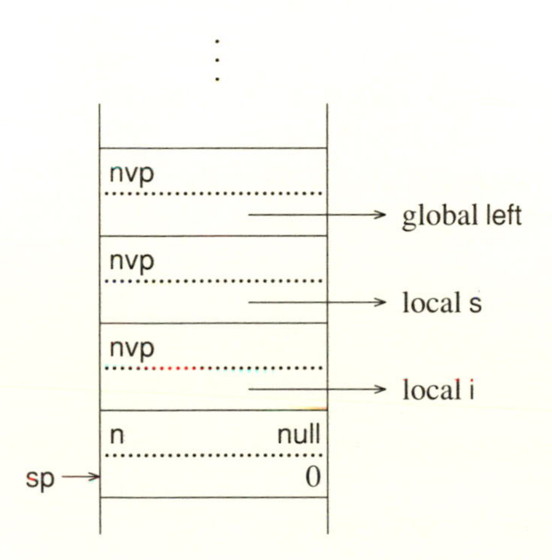


### 10.3.2 Function Invocation

Function invocation involves calling a C function in a fashion that is very similar to evaluating an operator. In the case of an Icon function, the entry point of the corresponding C function is obtained from the procedure block rather than by indexing an array of function pointers corresponding to operator codes.

For an Icon function that has a fixed number of arguments, the C function is called with a single argument that is a pointer to the location of Arg0 on the interpreter stack. Note that Arg0 is the descriptor that points to the procedure block. For an Icon function that may be called with an arbitrary number of arguments, the C function is called with two arguments: the number of arguments on the stack and a pointer to Arg0.

Like an operator, a function may fail, return a result, or suspend. The coding protocol is the same as for operators. The function find is an example:

```
FncDcl(find, 4)
   {
   register word l;
   register char *s1, *s2;
   word i, j, t;
   long l1, l2;
   char sbuf1[MaxCvtLen], sbuf2[MaxCvtLen];

   /*
    * Arg1 must be a string.  Arg2 defaults to &subject; Arg3 defaults
    *  to &pos if Arg2 is defaulted, or to 1 otherwise; Arg4 defaults
    *  to 0.

    */
   if (cvstr(&Arg1, sbuf1) == CvtFail)
      runerr(103, &Arg1);
   if (defstr(&Arg2, sbuf2, &k_subject))
      defint(&Arg3, &l1, k_pos);
   else
      defint(&Arg3, &l1, (word)1);
   defint(&Arg4, &l2, (word)0);

   /*
    * Convert Arg3 and Arg4 to absolute positions in Arg2 and order them.
    */
   i = cvpos(l1, StrLen(Arg2));
   if (i == 0)
      Fail;
   j = cvpos(l2, StrLen(Arg2));
   if (j == 0)
      Fail;
   if (i > j) {
      t = i;
      i = j;
      j = t;
      }
```

```
/*
 * Loop through Arg2[i:j] trying to find Arg1 at each point, stopping
 *  when the remaining portion Arg2[i:j] is too short to contain Arg1.
 */
Arg0.dword = D_Integer;
while (i <= j - StrLen(Arg1)) {
   s1 = StrLoc(Arg1);
   s2 = StrLoc(Arg2) + i - 1;
   l = StrLen(Arg1);

   /*
    * Compare strings on a byte-wise basis; if the end is reached
    *  before inequality is found, suspend with the position of the
    *  string.
    */
   do {
      if (l-- <= 0) {
         IntVal(Arg0) = i;
         Suspend;
         break;
         }
      } while (*s1++ == *s2++);
   i++;
   }

Fail;
}
```

### 10.3.3 Procedure Invocation

In the case of procedure invocation, a *procedure frame* is pushed onto the interpreter stack to preserve information that may be changed during the execution of the procedure and that must be restored when the procedure returns. As for other types of frames, a procedure frame begins with a marker. A procedure frame marker consists of eight words:

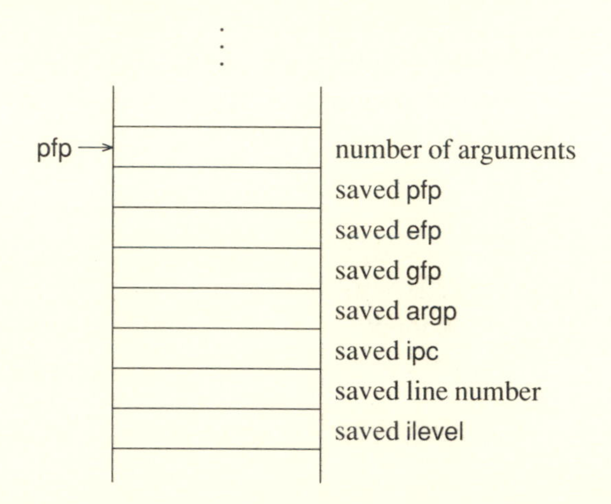


The current procedure frame is pointed to by pfp, and argp points to the place on the interpreter stack where the arguments begin, analogous to Arg0 for functions. The number of arguments, which can be computed from pfp and argp, is provided to make computations related to arguments more convenient.

After the procedure marker is constructed, a null-valued descriptor is pushed for each local identifier. For example, the call

```
calc(3,4)
```

for the procedure declaration given in Sec. 10.2 produces

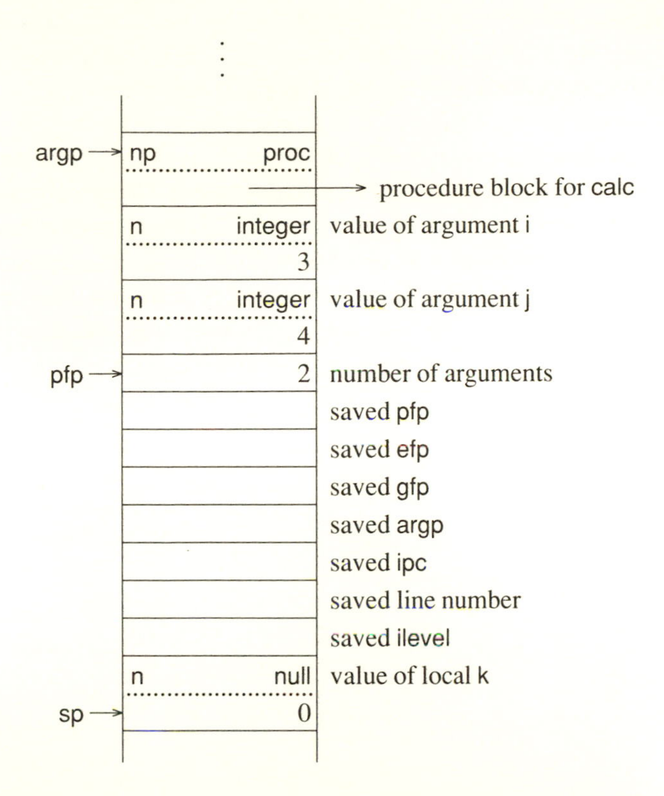


Once the null values for the local identifiers are pushed, ipc is set to the entry point given in the procedure block and efp and gfp are set to zero. Execution then continues in the interpreter with the new ipc.

The three forms of return from a procedure are the same as those from a function and correspond to the source-language expressions

```
return e
fail
suspend e
```

The corresponding virtual machine instructions are pret, pfail, and psusp. For example, the virtual machine code for

```
return &null
```

is

```
pnull
pret
```

In the case of pret, the result currently on the top of the interpreter stack is copied on top of the descriptor pointed to by argp. If this result is a variable that is on the stack (and hence local to the current procedure call), it is dereferenced in place. The C stack is unwound, since there may be suspended generators at the time of the return. The values saved in the procedure frame marker are restored, and execution continues in the interpreter with the restored ipc.

In the case of failure, the C stack is unwound as it is for pret, values are restored from the procedure frame marker, and control is transferred to efail.

Procedure suspension is similar to other forms of suspension. The descriptor on the top of the interpreter stack is dereferenced, if necessary, and saved. A generator frame marker is constructed on the interpreter stack to preserve values that may be needed if the procedure call is resumed. For procedure suspension, a generator frame marker contains two words in addition to those needed for other kinds of generator frame markers and has the form

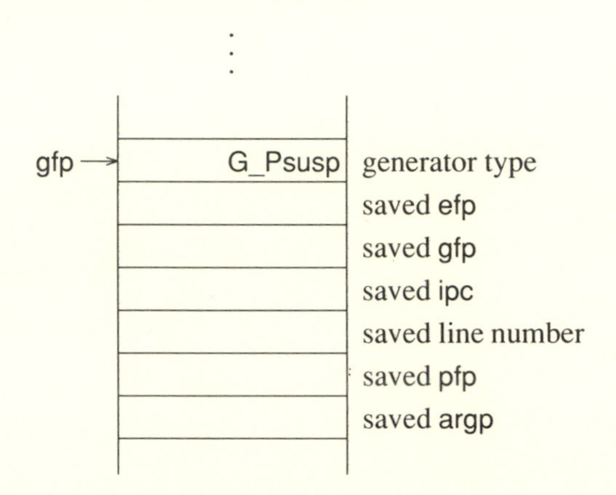


After the generator frame marker is pushed, the portion of the stack between the last generator or expression frame marker before the call to this procedure and the word prior to argp is copied to the top of the stack. Finally, the saved descriptor, which is the result produced by the procedure, is pushed on the top of the stack. Execution then continues in the interpreter with the restored ipc.

## 10.4 CO-EXPRESSIONS

Co-expressions add another dimension to expression evaluation in Icon. The important thing to understand about co-expressions is that Icon evaluation is always in *some* co-expression. Although it is not evident, the execution of an Icon program begins in a co-expression, namely the value of &main.

A co-expression requires both an interpreter stack and a C stack. In the co-expression for &main, the interpreter stack is statically allocated and the C stack is the one normally used for C execution—the ''system stack'' on some computers. The creation of a new co-expression produces a new interpreter stack and a new C stack, as well as space that is needed to save state information. When a co-expression is activated, the context for evaluation is changed to the stacks for the activated co-expression. When the activation of a co-expression produces a result, it in turn activates the co-expression that activated it, leaving the stacks from which the return occurred in a state of suspension. Thus, co-expression activation constitutes a simple context switch. In every co-expression, expression evaluation is in some state, possibly actively executing, possibly suspended, or possibly complete and unreachable.

The virtual machine instructions for

create *expr*₀

are

```
        goto      L3
L1:
        pop
        mark      L2
        code for expr0
        coret
        efail
L2:
        cofail
        goto      L2
L3:
        create    L1
```

Control goes immediately to L3, where the instruction create constructs a co-expression block and returns a descriptor that points to it. This block contains space for i-state variables, space for the state of the C stack, an interpreter stack, and a C stack.

The code between L1 and L3 is not executed until the co-expression is activated. The pop instruction following L1 discards the result transmitted to a co-expression on its first activation, since there is no expression waiting to receive the result of an initial activation. Next, an expression frame marker is created, and the code for $expr_0$ is executed. If $expr_0$ produces a result, coret is executed to return the result to the activating expression. If the co-expression is activated again, its execution continues with efail, which causes any suspended generators in the code for $expr_0$ to be resumed. If $expr_0$ fails, the expression frame is removed and cofail is executed. The cofail instruction is very similar to the coret instruction, except that it signals failure rather than producing a result. Note that if a co-expression that returns by means of cofail is activated again, the cofail instruction is executed in a loop.

A co-expression is activated by the expression

*$expr_1$ @ $expr_2$*

for which the virtual machine code is

```
code for expr1
code for expr2
coact
```

The more common form of activation, @$expr_0$, is just an abbreviation for &null @ $expr_0$; a result is always transmitted, even if it is the null value.

The virtual machine code for $expr_1$ produces the descriptor for the result that is to be transmitted to the co-expression being activated. The coact instruction dereferences the result produced by $expr_2$, if necessary, and checks to make sure it is a co-expression. After setting up state information, coact transfers control to the new co-expression with ipc set to L1. Execution continues there. If coret is reached, control is restored to the activating co-expression. The instructions coact and coret are very similar. Each saves the current co-expression state, sets up the new co-expression state, and transfers control.

**Co-Expression Blocks.** There is quite a bit of information associated with a co-expression, and space is provided for it in a co-expression block:

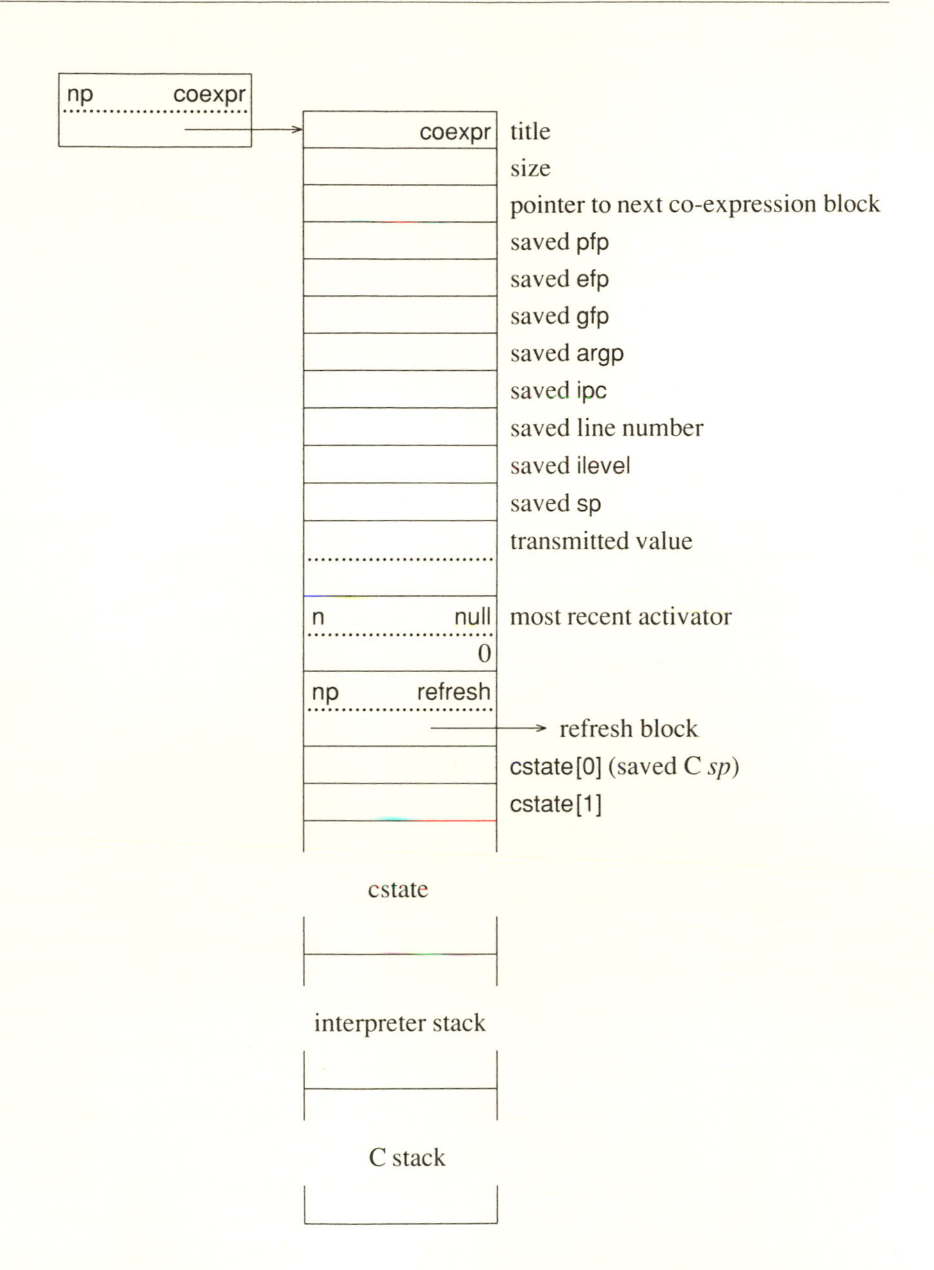


The interpreter stack and C stack shown in this diagram are not to scale compared with the rest of the block. Both are comparatively large; the actual sizes depend on the address space of the target computer.

The first word of the block is the usual title. The next word contains the number of results the co-expression has produced—its "size." Then there is a pointer to the next co-expression block on a list that is maintained for garbage-collection purposes. See Sec. 11.3.4. Following this pointer there are i-state variables: pfp, efp, gfp, argp, ipc, sp, the current program line number, and ilevel.

Then there is a descriptor for the transmitted result, followed by two more descriptors: one for the co-expression that activates this one and one for a *refresh* block that is needed if a copy of this co-expression block is needed. C state information is contained in an array of words, cstate, for registers and possibly other state information. The array cstate typically contains fifteen words for such information. The C *sp* is stored in cstate[0]. The use of the rest of cstate is machine-dependent.

Finally, there is an interpreter stack and a C stack. On a computer with a downward-growing C stack, such as the VAX, the base of the C stack is at the end of the co-expression block and the interpreter and C stacks grow toward each other. On a computer with an upward-growing C stack, the C stack base follows the end of the interpreter stack.

**Stack Initialization.** When a co-expression is first activated, its interpreter stack must be in an appropriate state. This initialization is done when the co-expression block is created. A procedure frame, which is a copy of the procedure frame for the procedure in which the create instruction is executed, is placed on the new stack. It consists of the words from argp through the procedure frame marker and the descriptors for the local identifiers. The efp and gfp in the co-expression block are set to zero and the ipc is set to the value given in the argument to the create instruction (L1).

No C state is set up on the new C stack; this is handled when the co-expression is activated the first time. The initial null value for the activator indicates the absence of a valid C state.

**Co-Expression Activation.** As mentioned previously, coact and coret perform many similar functions— both save current state information, establish new state information, and activate another co-expression. The current i-state variables are saved in the current co-expression block, and new ones are established from the co-expression block for the co-expression being activated. Similar actions are taken for the C state. Since the C state is machine-dependent, the ''context switch'' for the C state is performed by a routine, called coswitch, that contains assembly-language code.

The C state typically consists of registers that are used to address the C stack and registers that must be preserved across the call of a C function. On the VAX, for example, the C stack registers are *sp*, *ap*, and *fp*. Only the registers r6 through r11 must be saved for some C compilers, while other C compilers require that r3 through r11 be saved. Once the necessary registers are saved in the cstate array of the current co-expression, new values of these registers are established. If the co-expression being activated has been activated before, the C state is set up from its cstate array, and coswitch returns to interp. At this point, execution continues in the newly activated co-expression. Control is transferred to the beginning of the interpreter loop, and the next instruction (from the ipc for the co-expression) is fetched.

However, when a co-expression is activated for the first time, there are no register values to restore, since no C function has yet been called for the new co-expression. This is indicated, as mentioned previously, by a null activator, which is communicated to coswitch by an integer argument. In this case, coswitch sets up registers for the call of a C function and calls interp to start the execution of the new co-expression. Such a call to interp on the first activation of a co-expression corresponds to the call to interp that starts program execution in the co-expression &main for the main procedure. There can never be a return from the call to interp made in coswitch, since program execution can only terminate normally by a return from the main procedure, in &main.

The function coswitch is necessarily machine-dependent. The version for the VAX with the Berkeley 4.3bsd C compiler is an example:

```
coswitch(old_cs, new_cs, first)
int *old_cs, *new_cs;
int first;
   {
   asm("  movl 4(ap),r0");
   asm("  movl 8(ap),r1");
   asm("  movl sp,0(r0)");
   asm("  movl fp,4(r0)");
   asm("  movl ap,8(r0)");
   asm("  movl r11,16(r0)");
   asm("  movl r10,20(r0)");
   asm("  movl r9,24(r0)");
   asm("  movl r8,28(r0)");
   asm("  movl r7,32(r0)");
   asm("  movl r6,36(r0)");

   if (first == 0) {              /* this is the first activation */
      asm("  movl 0(r1),sp");
      asm("  clrl fp");
      asm("  clrl ap");
      interp(0, 0);
      syserr("interp() returned in coswitch");
      }
```

```
    else {
        asm(" movl 0(r1), sp");
        asm(" movl 4(r1), fp");
        asm(" movl 8(r1), ap");
        asm(" movl 16(r1), r11");
        asm(" movl 20(r1), r10");
        asm(" movl 24(r1), r9");
        asm(" movl 28(r1), r8");
        asm(" movl 32(r1), r7");
        asm(" movl 36(r1), r6");
        }
    }
```

The variables old_cs and new_cs are pointers to the cstate arrays for the activating and activated co-expressions, respectively. The value of first is 0 if the co-expression is being activated for the first time. Note that in order to write coswitch it is necessary to know how the first two arguments are accessed in assembly language. For the previous example, old_cs and new_cs are four and eight bytes from the *ap* register, respectively.

**Refreshing a Co-Expression.** The operation ^$expr_0$ creates a copy of the co-expression produced by $expr_0$ with its state initialized to what it was when it was originally created. The refresh block for $expr_0$ contains the information necessary to reproduce the initial state. The refresh block contains the original ipc for the co-expression, the number of local identifiers for the procedure in which $expr_0$ was created, a copy of the procedure frame marker at the time of creation, and the values of the arguments and local identifiers at the time of creation. Consider, for example,

```
procedure labgen(s)
    local i, j, e
    i := 1
    j := 100
    e := create (s || (i to j) || ":")
          :
          :
end
```

For the call labgen("L"), the refresh block for e is

| | |
|---|---|
| refresh | title |
| 88 | size of block |
| | initial ipc |
| 3 | number of local identifiers |
| 1 | number of arguments |
| | saved pfp |
| | saved efp |
| | saved gfp |
| | saved argp |
| | saved ipc |
| | saved line number |
| | saved ilevel |
| n proc | value of labgen |
| → | procedure block |
| 1 | value of s |
| → | "L" |
| n integer | value of i |
| 1 | |
| n integer | value of j |
| 100 | |
| n null | value of e |
| 0 | |

RETROSPECTIVE: Invocation expressions are more complicated to implement than operators, since the meaning of an invocation expression is not known until it is evaluated. Since functions and procedures are source-language values, the information associated with them is stored in blocks in the same manner as for other types.

The C code that implements Icon functions is written in the same fashion as the code for operators. Procedures have source-language analogs of the failure and suspension mechanisms used for implementing functions and operators. Procedure frames identify the portions of the interpreter stack associated with the procedures currently invoked.

A co-expression allows an expression to be evaluated outside its lexical site in the program by providing separate stacks for its evaluation. The possibility of multiple stacks in various states of evaluation introduces technical problems into the implementation, including a machine-dependent context switch.

## EXERCISES

**10.1** What happens if a call of a procedure or function contains an extra argument expression, but the evaluation of that expression fails?

**10.2** Sometimes it is useful to be able to specify a function or procedure by means of its string name. Icon supports "string invocation," which allows the value of $expr_0$ in

$expr_0(expr_1, expr_2, ..., expr_n)$

to be a string. Thus,

```
"write"(s)
```

produces the same result as

```
write(s)
```

Of course, such a string name is usually computed, as in

```
(read())(s)
```

Describe what is involved in implementing this aspect of invocation.

Operators also may be invoked by their string names, as in

```
"+"(i, j)
```

What is needed in the implementation to support this facility? Can a control structure be invoked by a string name?

**10.3** If the result returned by a procedure is a variable, it may need to be dereferenced. This is done in the code for pret and psusp. For example, if the result being returned is a local identifier, it must be replaced by its value. What other kinds of variables must be dereferenced? Is there any difference in the dereferencing done by pret and psusp?

**10.4** How is the structure of a co-expression block different on a computer with an upward-growing C stack compared to one with a downward-growing C stack? What is the difference between the two cases in terms of potential storage fragmentation?

CHAPTER 11

# Storage Management

PERSPECTIVE: The implementation of storage management must accommodate a wide range of allocation requirements. At the same time, the implementation must provide generality and some compromise between ''normal'' programs and those that have unusual requirements. Although it is clearly sensible to satisfy the needs of most programs in an efficient manner, there is no way to define what is typical or to predict how programming style and applications may change. Indeed, the performance of the implementation may affect both programming style and applications.

Strings and blocks can be created during program execution at times that cannot be predicted, in general, from examination of the text of a program. The sizes of strings and of some types of blocks may vary and may be arbitrarily large, although practical considerations dictate some limits. There may be an arbitrary number of strings and blocks. The ''lifetimes'' during which they may be used are arbitrary and are unrelated, in general, to procedure calls and returns.

Different programs vary considerably in the number, type, and sizes of data objects that are created at run time. Some programs read in strings, transform them, and write them out without ever creating other types of objects. Other programs create and manipulate many lists, sets, and tables but use few strings. Relatively few programs use co-expressions, but there are applications in which large numbers of co-expressions are created.

Since a program can construct an arbitrary number of data objects of arbitrary sizes and lifetimes, some mechanism is needed to allow the reuse of space for ''dead'' objects that are no longer accessible to the program. Thus, in addition to a mechanism for allocating storage for objects at run time, there must be a storage-reclamation mechanism, which usually is called *garbage collection*. The methods used for allocation and garbage collection are interdependent. Simple and fast allocation methods usually require complex and time-consuming garbage-collection techniques, while more efficient garbage-collection techniques generally lead to more complex allocation techniques.

Storage management has influences that are far-reaching. In some programs, it may account for a major portion of execution time. The design of data structures, the layout of blocks, and the representation of strings are all influenced by storage-management considerations. For example, both a descriptor that points to a block and the first word of the block contain the same type code. This information is redundant as far as program execution is concerned, since blocks are accessed only via descriptors that point to them. The redundant type information is motivated by storage-management considerations. During garbage collection, it is necessary to access blocks directly, rather than through pointers from descriptors, and it must be possible to determine the type of a block from the block itself. Similarly, the size of a block is of no interest in performing language operations, but the size is needed during garbage collection. Blocks, therefore, carry some ''overhead'' for storage management. This overhead consists primarily of extra space, reflecting the fact that it takes more space to manage storage dynamically than would be needed if space were allocated statically. Balancing space overhead against efficiency in allocating and collecting objects is a complex task.

Such problems have plagued and intrigued implementors since the early days of LISP. Many ways have been devised to handle dynamic storage management, and some techniques have been highly refined to meet specific requirements (Cohen 1981). In the case of Icon, there is more emphasis on storage management for strings than there is in a language, such as LISP, where lists predominate. Icon's storage-management system reflects previous experience with storage-management systems used in XPL (McKeeman, Horning, and Wortman 1970), SNOBOL4 (Hanson 1977), and the earlier Ratfor implementation of Icon (Hanson 1980). The result is, of course, somewhat idiosyncratic, but it provides an interesting case study of a real storage-management system.

## 11.1 MEMORY LAYOUT

During the execution of an Icon program, memory is divided into several regions. The sizes and locations of these regions are somewhat dependent on computer architecture and the operating system used, but typically they have the following form:

| run-time system |
| --- |
| icode |
| allocated storage |
| free space |
| system stack |

This diagram is not drawn to scale; some regions are much larger than others.

**The Run-Time System.** The run-time system contains the executable code for the interpreter, built-in operators and functions, support routines, and so forth. It also contains static storage for some Icon strings and blocks that appear in C functions. For example, the blocks for keyword trapped variables are statically allocated in the data area of the run-time system. Such blocks never move, but their contents may change. The size of the run-time system is somewhat machine-dependent. About 75,000 bytes (decimal) is typical.

**The Icode Region.** One of the first things done by the run-time system is to read in the icode file for the program that is to be executed. The data in the icode region, which is produced by the linker, is divided into a number of sections:

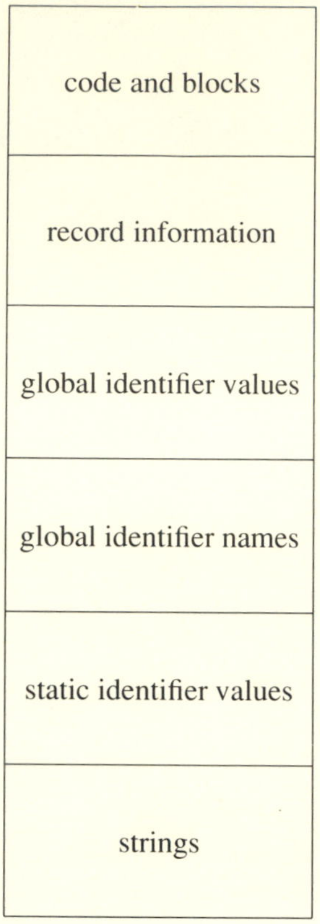

| code and blocks |
| --- |
| record information |
| global identifier values |
| global identifier names |
| static identifier values |
| strings |

The first section contains virtual machine code, blocks for cset and real literals, and procedure blocks, on a per-procedure basis. Thus, the section of the icode region that contains code and blocks consists of segments of the following form for each procedure:

| blocks for real literals |
|---|
| blocks for cset literals |
| procedure block |
| virtual machine instructions |

Record information for the entire program is in the second section of the icode region. Next, there is an array of descriptors for the values of the global identifiers in the program, followed by an array that contains qualifiers for the names of the global identifiers. These two arrays are parallel. The *i*th descriptor in the first array contains the value of the *i*th global identifier, and the *i*th descriptor in the second array contains a qualifier for its name.

Following the two arrays related to global identifiers is an array for the values of static identifiers. As mentioned in Sec. 2.1.10, static identifiers have global scope with respect to procedure invocation, but a static identifier is accessible only to the procedure in which it is declared.

Unlike cset and real blocks, which are compiled on a per-procedure basis, all strings in a program are pooled and are in a single section of the icode region that follows the array of static identifiers. A literal string that occurs more than once in a program occurs only once in the string section of the icode region.

Data in the icode region is never moved, although some components of it may change at run time. The size of the icode region depends primarily on the size of the corresponding source program. As a rule of thumb, an icode region is approximately twice as large as the corresponding Icon source-language file. An icode file for a short program might be 1,000 bytes, while one for a large program (by Icon standards) might be 20,000 bytes.

**Allocated Storage.** The space for data objects that are constructed at run time is provided in allocated storage regions. This portion of memory is divided into three parts:

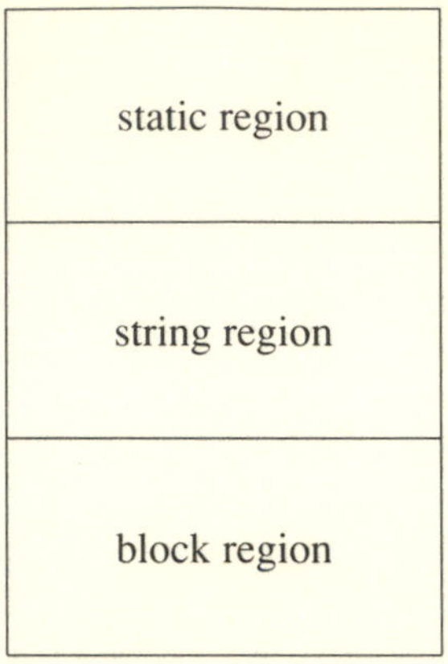

| static region |
|---|
| string region |
| block region |

The static region contains co-expression blocks. The remainder of the allocated storage region is divided into strings and blocks as shown. The string region contains only characters. The block region, on the other hand, contains pointers. This leads to a number of differences in allocation and garbage-collection techniques in different regions.

Data in the static region is never moved, but strings and blocks may be. Both the string and block regions may be moved if it is necessary to increase the size of the static region. Similarly, the block region may be moved in order to enlarge the string region.

The initial sizes of the allocated storage regions vary considerably from computer to computer, depending on the size of the user address space. On a computer with a large address space, such as the VAX, the initial sizes are

| | | |
|---|---|---|
| static region: | 20,480 bytes | (5,120 words) |
| string region: | 51,200 bytes | (12,800 words) |
| block region: | 51,200 bytes | (12,800 words) |
| total: | 122,880 bytes | (30,720 words) |

On a computer with a small address space, such as the PDP-11, the initial sizes are

| | | |
|---|---|---|
| static region: | 4,096 bytes | (2,048 words) |
| string region: | 10,240 bytes | (5,120 words) |
| block region: | 10,240 bytes | (5,120 words) |
| total: | 24,576 bytes | (12,288 words) |

The user may establish different initial sizes prior to program execution by using environment variables. As indicated previously, the sizes of these regions are increased at run time if necessary, but there is no provision for decreasing the size of a region once it has been established.

**Free Space and the System Stack.** On computers with system stacks that grow downward, such as the VAX, the system stack grows toward the allocated storage region. Between the two regions is a region of free space into which the allocated storage region may grow upward. Excessive recursion in C may cause collision of the system stack and the allocated storage region. This is an unrecoverable condition, and the result is termination of program execution. Similarly, more space may be needed for allocated storage than is available. This also results in termination of program execution. In practice, the actual situation depends to a large extent on the size of the user address space, which is the total amount of memory that is available for all the regions shown previously. On a computer with a small user address space, such as the PDP-11, the amount of memory available for allocated storage is a limiting factor for some programs. Furthermore, collision of the allocated storage region and the system stack is a serious problem. On a computer that supports a large virtual memory, the size of the system stack is deliberately limited, since the the total amount of memory available is so large that runaway recursion would consume enormous resources before a collision occurred between the system stack and the allocated storage region.

## 11.2 ALLOCATION

Storage allocation in Icon is designed to be fast and simple. Garbage collection is somewhat more complicated as a result. Part of the rationale for this approach is that most Icon programs do a considerable amount of allocation, but many programs never do a garbage collection. Hence, programs that do not garbage collect are not penalized by a strategy that makes garbage collection more efficient at the expense of making allocation less efficient. The other rationale for this approach is that the storage requirements of Icon do not readily lend themselves to more complex allocation strategies.

### 11.2.1 The Static Region

Data allocated in the static region is never moved, although it may be freed for reuse. Co-expression blocks are allocated in the static region, since their C stacks contain internal pointers that depend on both the computer and the C compiler and hence are difficult to relocate to another place in memory. Furthermore, since co-expression blocks are all the same size, it is economical and simple to free and reuse their space.

The C library routines malloc and free are used to allocate and free co-expression blocks in the static region. These routines maintain a list of blocks of free space. The routine malloc finds a block of the requested size, dividing a larger block if necessary, and revises the free list accordingly. The routine free returns the freed space to the free list, coalescing it with adjacent free blocks if possible. See Kernighan and Ritchie 1978 for a discussion of free-list allocation.

Icon contains its own version of these routines to assure that space is allocated in its own static region and to allow its overall memory region to be expanded without conflict with other users of malloc. Thus, if a user extension to Icon or the operating system calls malloc, Icon's own routine handles the request. This means that the static region may contain space allocated for data other than co-expression blocks, although this normally is not the case.

### 11.2.2 Blocks

For other kinds of blocks, Icon takes advantage of the fact that its own data can be relocated if necessary and uses a very simple allocation technique. The allocated region for blocks is divided into two parts:

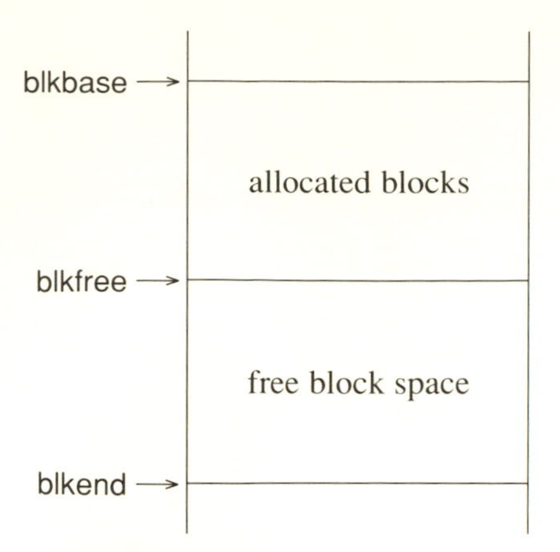


When there is a request for a block of $n$ bytes, the free pointer, blkfree, is incremented by $n$ and the previous value of the free pointer is returned as the location of the newly allocated block. This process is fast and free of the complexities of the free-list approach.

Note that this technique really amounts to a free list with only one block. The problem of reclaiming fragmented space on the free list is exchanged for the process of reclaiming unused blocks and rearranging the block region so that all the free space is in one contiguous portion of the block region. This is done during garbage collection.

### 11.2.3 Strings

There is even less justification for a free-list approach for allocating strings. A newly created string may be one character long or it may be thousands of characters long. Furthermore, while there is space in blocks that can be used to link together free storage, there is no such space in strings, and a free list would involve additional storage.

Instead, the string region is allocated in the same way that the block region is allocated:

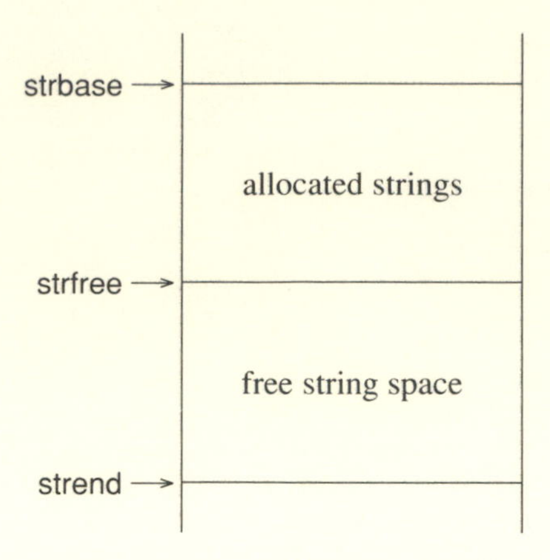


As with the block region, a garbage collection is performed if there is not enough space in the string region to satisfy an allocation request.

## 11.3 GARBAGE COLLECTION

Allocation is simple, but garbage collection is not. The primary purpose of garbage collection is to reclaim the space occupied by ''dead'' objects that are not needed for subsequent program execution, so that this space can be reallocated. This means different things in different regions. In the static region, it means freeing dead co-expression blocks. In the string and block regions, it involves moving the space for dead objects from the allocated portion of the region to the free portion. This is considerably more complicated than adding a pointer to a free list. Since all free space must be in a single block in these regions, ''live'' objects must be moved to fill in the holes left by dead ones. This is done by compacting the allocated portion of these regions, relocating live objects toward the beginning of these regions and squeezing out dead objects. In turn, pointers to live objects have to be adjusted to correspond to their new locations. There are two phases in garbage collection:

- Location of live objects and all the pointers to them.
- Compaction of live objects and adjustment of the pointers to them.

''Garbage collection'' is somewhat of a misnomer, since the process is oriented toward saving ''non-garbage'' objects; garbage disappears as a byproduct of this operation.

### 11.3.1 The Basis

The challenging problem for garbage collection is the location of objects that have to be saved, as well as all pointers to them. An object is dead, by definition, if it cannot be accessed by any future source-language computation. Conversely, by definition, an object is live if it can be accessed. Consequently, the important issue is the possibility of computational access. For example, it is always possible to access the value of &subject, and this value must be preserved by garbage collection. On the other hand, in

```
a := [1,2,3]
a := list(10)
```

after the execution of the second assignment, the first list assigned to a is inaccessible and can be collected.

It is essential to save any object that may be accessed, but there is no way, in general, to know if a specific object *will* be accessed. For example, a computational path may depend on factors that are external to the program, such as the value of data that is read from a file. It does comparatively little harm to save an object that might be accessed but, in fact, never is. Some storage is wasted, but it is likely to be reclaimed during a subsequent collection. It is a serious error, on the other hand, to discard an object that subsequently *is* accessed. In the first place, the former value of such an object usually is overwritten and hence is ''garbage'' if it is subsequently accessed. Furthermore, accessing such an object may overwrite another accessible object that now occupies the space for the former one. The effects may range from incorrect computational results to addressing violations. The sources of such errors also are hard to locate, since they may not be manifested until considerably later during execution and in a context that is unrelated to the real cause of the problem. Consequently, it is important to be conservative and to err, if at all, on the side of saving objects whose subsequent accessibility is questionable. Note that it is not only necessary to locate all accessible objects, but it is also necessary to locate all pointers to objects that may be relocated.

The location phase starts with a *basis* that consists of descriptors that point to objects that may be accessible and from which other objects may be accessed. For example, &subject is in the basis. The precise content of the basis is partly a consequence of properties of the Icon language and partly a consequence of the way the run-time system is implemented. The basis consists of the following descriptors:

- &main (co-expression block for the initial call of main)
- current co-expression block
- values of global identifiers
- values of static identifiers
- &subject
- saved values of map arguments
- tended descriptors

The tended descriptors provide temporary storage for a run-time support routine in which a garbage collection may occur. See Sec 12.2.2.

Not all objects that have to be saved are pointed to directly by the basis. The value of a local identifier on the interpreter stack may point to a list-header block that in turn points to a list-element block that contains elements pointing to strings and other blocks. Pointer chains also can be circular.

### 11.3.2 The Location Phase

For historical reasons, the location phase is sometimes called *marking*. This term refers to the common practice of setting an identifying bit in objects that have been located. Not all such processes actually change the objects that are located. The way that this is done in Icon depends on the region in which an object is located.

During the location phase, every descriptor in the basis is examined. A descriptor is of interest only if it is a qualifier or if its v-word contains a pointer (that is, if its d-word contains a p flag). For a pointer dp to a descriptor, the following checks are performed:

```
if (Qual(*dp))
   postqual(dp);
else if (Pointer(*dp))
   markblock(dp);
```

where the macro Pointer(d) tests the d-word of d for a p flag.

**Strings.** The routine postqual first checks that the v-word of the qualifier points to a string in the allocated string region, since strings in other parts of memory are not of interest during garbage collection. If the string is in the allocated string region, a pointer to the qualifier is placed in an array:

```
postqual(dp)
struct descrip *dp;
   {
         .
         .
         .
   if (StrLoc(*dp) >= strbase && StrLoc(*dp) < strend)
      *qualfree++ = dp;
   }
```

The array quallist is empty when garbage collection begins. Its size is checked before a pointer is added to it, and more space is obtained if it is needed, although the code for doing that is not shown here. See Sec. 11.3.6.

The pointers that accumulate in quallist during the marking phase provide the information necessary to determine the portion of the allocated string region that is in use. In addition, these pointers point to all the qualifiers whose v-words must be adjusted when the strings they point to are moved during the compaction of string region. See Sec. 11.3.3.

**Blocks.** The location phase for blocks is more complicated than that for strings, since blocks can contain descriptors that point to strings as well as to other blocks. The objects that these descriptors point to must be processed also.

Unlike strings, in which a separate array is used to keep track of qualifiers that have been located, no extra space is needed to keep track of descriptors that point to blocks. Instead, descriptors and the titles of the blocks they point to are modified temporarily.

The title of any block located in the allocated block region is changed to point to a *back chain* that contains all the descriptors that point to that block. The descriptors are linked through their v-words.

The following example illustrates the process. Suppose there is a record declaration

```
record term(value, code, count)
```

and that the following expressions are evaluated:

```
x := term("chair", "noun",4)
y := x
```

The values of x, y, and the block they point to are related as follows:

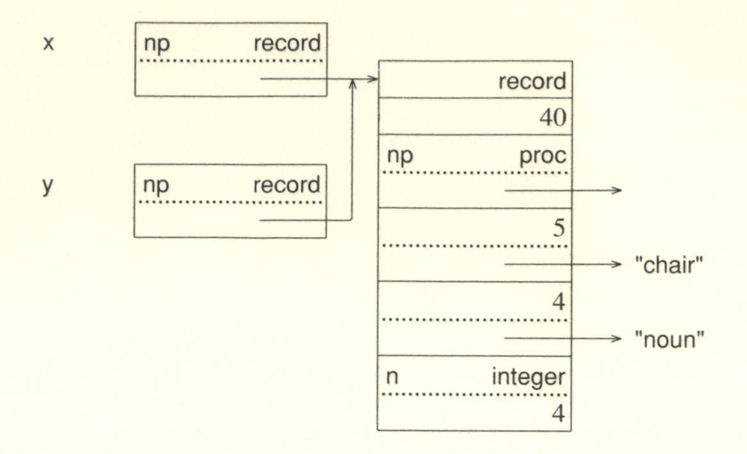


Suppose that the descriptor containing the value of x is processed during the location phase before the descriptor containing the value of y. This descriptor is identified as pointing to a block in the allocated block region by virtue of the p flag in its d-word and an address range check on the value of its v-word. The back chain is established by exchanging the contents of the title word of the block with the v-word of the descriptor that points to it. The result is

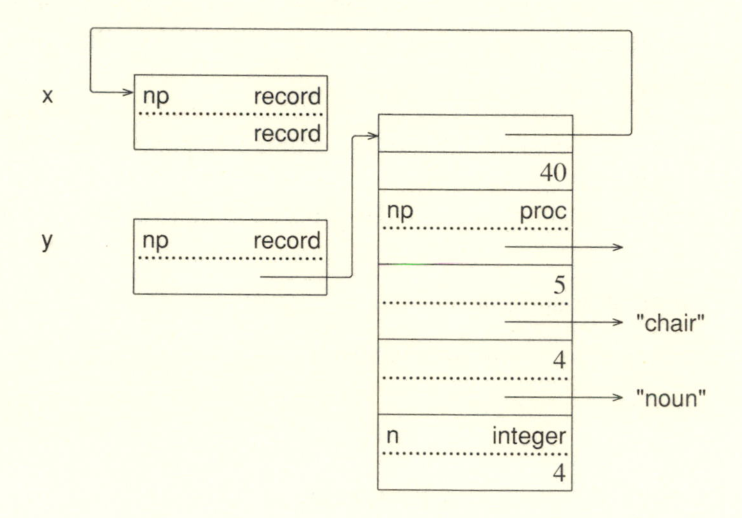


The title word of the block now points to the descriptor that previously pointed to the block. This change is reversible, and prior to the completion of the garbage-collection process the previous relationship is restored. A crucial but somewhat subtle aspect of the change is that it is now possible to tell that the block has been marked. The numerical magnitude of the value of its title word is greater than that of any type code, since all descriptors in the run-time system are at memory locations whose addresses are larger than the largest type code.

The descriptors in the record block now are processed in the same way as descriptors in the basis. In order to do this, it is necessary to know where descriptors are located in the block. Since blocks in the allocated block region are organized so that all descriptors follow all non-descriptor data, it is only necessary to know where the first descriptor is and how large the block is. These values are determined using two arrays that have entries for each type code.

The first array, bsizes, provides the information that is needed to determine block sizes. There are three kinds of entries. An entry of −1 indicates a type for which there is no block or for which the blocks are not in the allocated block region. Examples are T_Null and T_Coexpr. An entry of 0 indicates that the size of the block follows the block title. This is the case for records. Any other entry is the actual size of the block in bytes. For example, the entry in bsizes for T_List is 24 on a 32-bit computer.

The second array, firstd, is used to determine the byte offset of the first descriptor in the block. As with bsizes, a value of −1 indicates a type for which there are no associated blocks in the allocated block region. A value of 0 indicates that there are no descriptors in the block. Examples are T_Cset and T_Real. For T_Record, the entry is 8 for 32-bit computers, indicating that the first descriptor is at an offset of 8 bytes (2 words) from the beginning of the block. See Sec. 4.2.

For the previous example, after the descriptors in the record block are processed, the location phase continues. When the descriptor that contains the value of y is processed, it is added to the back chain by again exchanging the contents of its v-word with the contents of the title of the block. As a result, the title of the block points to the descriptor for the value of y and its v-word points to the descriptor for the value of x:

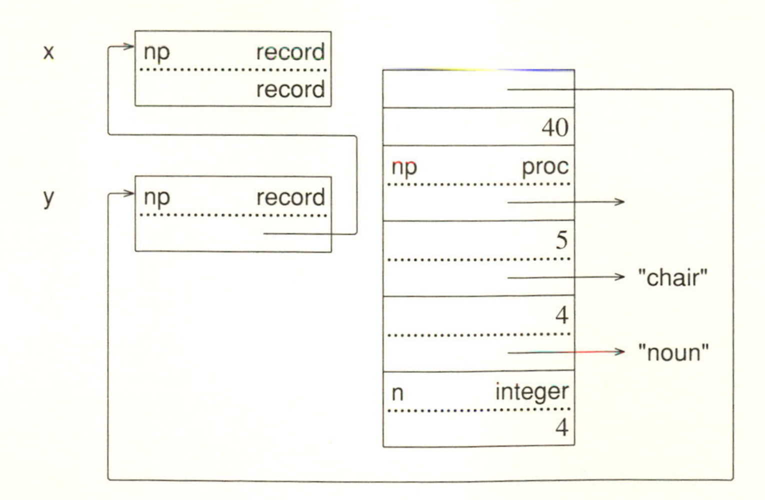


Since the title of the block that y points to is marked, the descriptors in it are not processed. This prevents descriptors from being processed twice and also prevents the marking phase from looping in case there are pointer loops among blocks.

If a variable descriptor is encountered when processing descriptors whose d-words contain p flags, the value the variable points to belongs to one of the following categories:

- trapped-variable block
- global or static identifier
- argument or local identifier
- descriptor in a structure

A trapped variable, indicated by a t flag in its v-word, points to a block and is processed like any other descriptor that points to a block. The values of global and static identifiers are in the basis and are processed separately. The values of arguments and local identifiers are on an interpreter stack and are processed when its co-expression block is processed. A variable descriptor that points to a descriptor in a structure points *within* a block, not to the title of a block. This is the only case in which the offset, which is contained in the least-significant portion of the d-word of a non-trapped-variable descriptor, is nonzero. Consequently, this offset is used to distinguish such variables from those in the second and third categories.

Continuing the previous example, suppose that a garbage collection is triggered by evaluation of the expression

```
x.count := read()
```

At the beginning of garbage collection, there is a variable descriptor for the field reference that points to the record block in addition to the descriptors for the values of x and y. If the values of x and y are processed first as described previously, the situation when the variable descriptor is encountered is

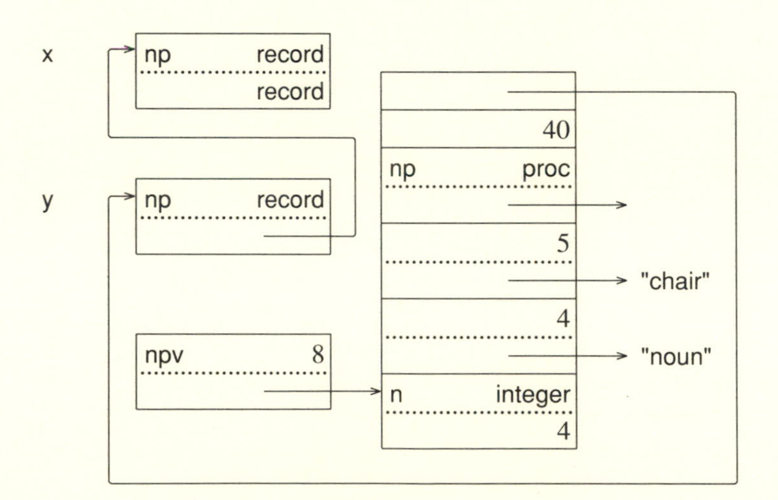


Note that the offset in the d-word of the variable descriptor is in words, not bytes.

The offset, converted to bytes, is added to the v-word in the variable descriptor, and this descriptor is linked into the back chain.

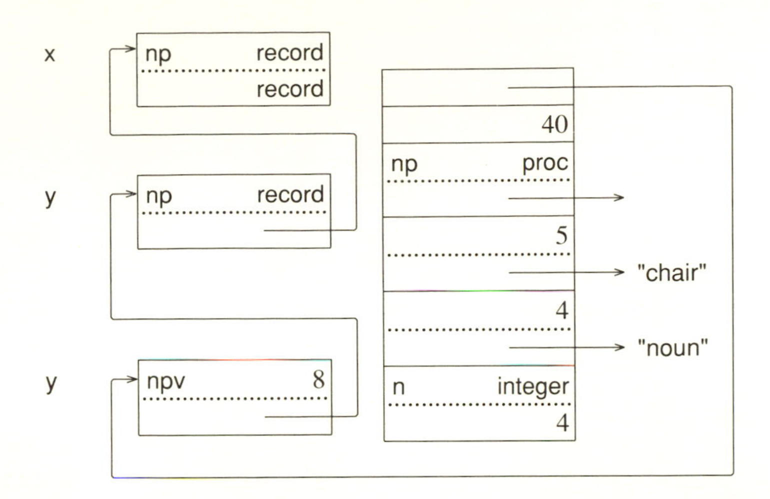


When the location phase is complete, the title of each block in the allocated block region that must be saved points to a chain of all the descriptors that originally pointed to it. This provides the necessary information to adjust the v-words of these descriptors to account for the relocation of the block during the compaction phase. See Sec. 11.3.3.

If a descriptor that points to a co-expression block is encountered during the location phase, the title of the co-expression block is marked and the descriptors in the co-expression block are processed in a fashion similar to that for blocks in the allocated block region. Since co-expression blocks are never moved, it is not necessary to keep track of descriptors that point to them. To mark the title, it is sufficient to change it to a value that is larger than any type code.

The activator of the co-expression (if any) is processed like any other co-expression block. Similarly, the refresh block that is pointed to from the co-expression block must be processed like any other block. The rest of the descriptors associated with a co-expression are in its interpreter stack.

Here the situation is more complicated than it is with blocks in the allocated block region, since interpreter stacks contain frame markers in addition to descriptors. Despite this, all the descriptors, and only the descriptors, on an interpreter stack must be processed.

Interpreter stacks are processed by the routine sweep, which starts at sp for the stack and works toward the stack base. Descriptors are processed until the next frame marker is encountered, at which point, depending on the type of the frame, the marker is skipped and new frame pointers are set up from it.

The routine for marking blocks is

```
markblock(dp)
struct descrip *dp;
   {
   register struct descrip *dp1;
   register char *endblock, *block;
   static word type, fdesc, off;

   /*
    * Get the block to which dp points.
    */

   block = (char *)BlkLoc(*dp);
   if (block >= blkbase && block < blkfree) {  /* check range */
      if (Var(*dp) && !Tvar(*dp)) {

         /*
          * The descriptor is a variable; point block to the head of the
          *  block containing the descriptor to which dp points.
          */
         off = Offset(*dp);
         if (off == 0)
            return;
         else
            block = (char *)((word *)block - off);
            }

         type = BlkType(block);
         if ((uword)type <= MaxType)  {

            /*
             * The type is valid, which indicates that this block has not
             *  been marked.  Point endblock to the byte past the end
             *  of the block.
             */
            endblock = block + BlkSize(block);
            }

         /*
          * Add dp to the back chain for the block and point the
          *  block (via the type field) to dp.
          */
         BlkLoc(*dp) = (union block *)type;
         BlkType(block) = (word)dp;
         if (((unsigned)type <=  MaxType) && ((fdesc = firstd[type]) > 0))
```

```
                /*
                 * The block has not been marked, and it does contain
                 *  descriptors. Mark each descriptor.
                 */
                for (dp1 = (struct descrip *)(block + fdesc);
                   (char *) dp1 < endblock; dp1++) {
                   if (Qual(*dp1))
                      postqual(dp1);
                   else if (Pointer(*dp1))
                      markblock(dp1);
                }
             }
          else if (dp->dword == D_Coexpr &&
             (unsigned)BlkType(block) <= MaxType) {

             /*
              * dp points to a co-expression block that has not been
              *  marked. Point the block to dp. Sweep the interpreter
              *  stack in the block. Then mark the block for the
              *  activating co-expression and the refresh block.
              */
             BlkType(block) = (word)dp;
             sweep((struct b_coexpr *)block);
             markblock(&((struct b_coexpr *)block)->activator);
             markblock(&((struct b_coexpr *)block)->freshblk);
             }
       }
```

The macro BlkType(cp) produces the type code of the block pointed to by cp. The macro BlkSize(cp) uses the array bsizes to determine the size of the block pointed to by cp.

### 11.3.3 Pointer Adjustment and Compaction

**Strings.** When the location phase is complete, quallist contains a list of pointers to all the qualifiers whose v-words point to the allocated string region. For example, suppose that the allocated string region at the beginning of a garbage collection is

```
... Necessity is the mother of strange bedfellows ...
↑             ↑                                    ↑
strbase       +400                                 strfree
```

Suppose also that the following qualifiers reference the allocated string region:

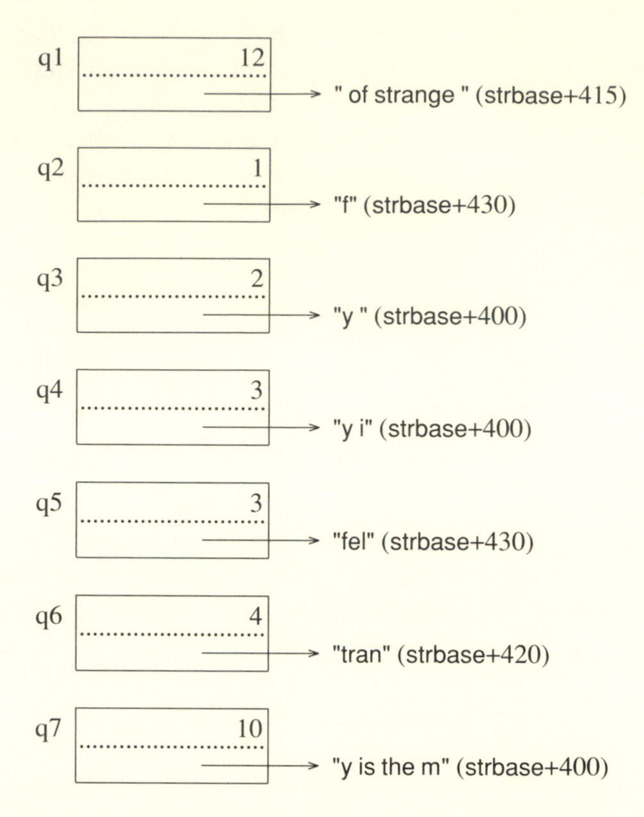


The pointers to the allocated string region are

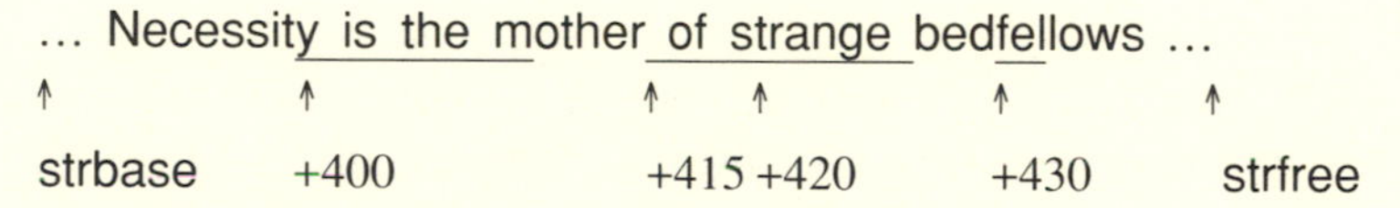


Note that the qualifiers point to overlapping strings.

After the location phase, quallist might contain the following pointers:

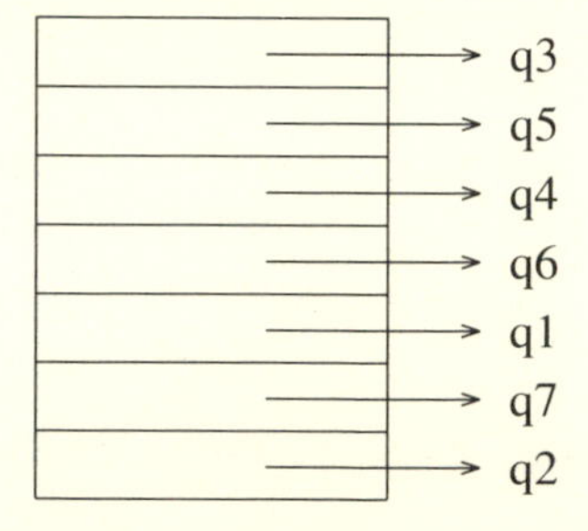


The order of the pointers in quallist depends on the order in which the qualifiers are processed; there is no necessary relationship between the order of the pointers in quallist and the order of the pointers to the allocated string region.

At the beginning of the pointer-adjustment phase of garbage collection, the array quallist is sorted in non-decreasing order by the v-words in qualifiers that are pointed to from quallist. This allows the pointers to the allocated string region to be processed in non-decreasing order so that the portions of the allocated string region that must be saved and compacted can be determined.

Continuing the previous example, quallist becomes

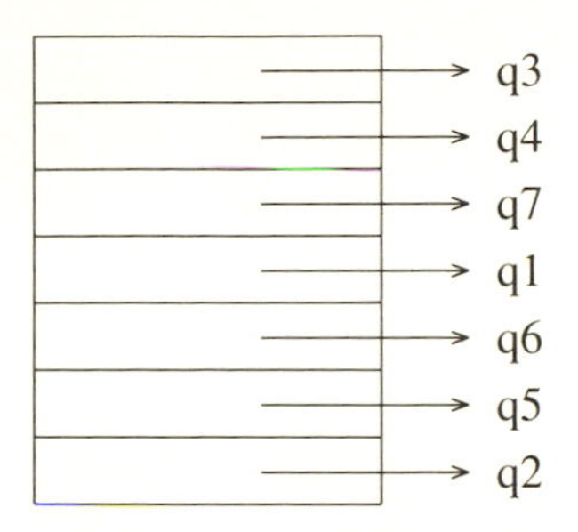


The v-words of the qualifiers in the order of the pointers in quallist now are

strbase+400
strbase+400
strbase+400
strbase+415
strbase+420
strbase+430
strbase+430

Since qualifiers may reference overlapping strings, care must be taken to identify contiguous "clumps" of characters that may be shared by qualifiers. The pointers in quallist are processed in order. Three pointers in the string region are maintained: dest, the next free destination for a clump of characters to be saved; source, the start of the next clump; and cend, the end character in the current clump.

When a qualifier that is pointed to from quallist is processed, the first question is whether its v-word addresses a character that is beyond the end of the current clump (since v-words are processed in numerical order, the address is either in the current clump or beyond the end of it). If it is in the current clump, cend is updated, provided the last character of the current qualifier is beyond cend. If it is not in the current clump, the clump is moved from source to dest. In either case, the v-word of the current qualifier is adjusted (dest – source is added to it).

In the previous example, the allocated string region after collection is

```
y is the m of strange fel
↑                     ↑
strbase               strfree
```

and the seven qualifiers that point to it are

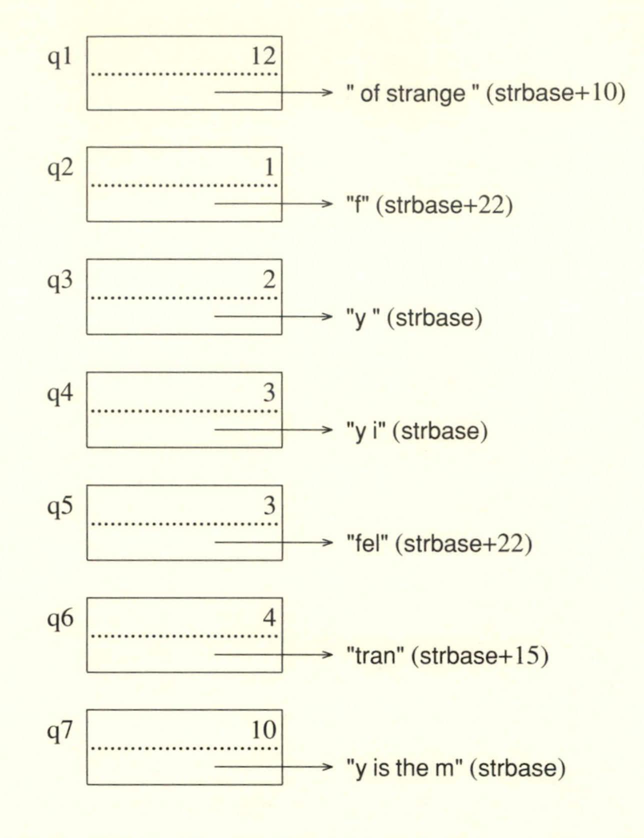


The routine for compacting the allocated string region and adjusting pointers to it is

```
scollect(extra)
word extra;
   {
   register char *source, *dest;
   register struct descrip **qptr;
   char *cend;
   extern int qlcmp();
```

```
if (qualfree <= quallist) {
   /*
    * There are no accessible strings. Thus, there are none to
    *  collect and the whole string space is free.
    */
   strfree = strbase;
   return;
   }
/*
 * Sort the pointers on quallist in ascending order of string
 *  locations.
 */
qsort(quallist, qualfree - quallist, sizeof(struct descrip *), qlcmp);
/*
 * The string qualifiers are now ordered by starting location.
 */
dest = strbase;
source = cend = StrLoc(**quallist);

/*
 * Loop through qualifiers for accessible strings.
 */
for (qptr = quallist; qptr < qualfree; qptr++) {
   if (StrLoc(**qptr) > cend) {

      /*
       * qptr points to a qualifier for a string in the next clump.
       *  The last clump is moved, and source and cend are set for
       *  the next clump.
       */
      while (source < cend)
         *dest++ = *source++;
      source = cend = StrLoc(**qptr);
      }
```

```
        if (StrLoc(**qptr)+StrLen(**qptr) > cend)
           /*
            * qptr is a qualifier for a string in this clump; extend
            *  the clump.
            */
           cend = StrLoc(**qptr) + StrLen(**qptr);
        /*
         * Relocate the string qualifier.
         */
        StrLoc(**qptr) += dest - source + extra;
        }

     /*
      * Move the last clump.
      */
     while (source < cend)
        *dest++ = *source++;
     strfree = dest;
     }
```

The argument extra provides an offset in case the string region is moved. See Sec. 11.3.5.

Sorting is done by the C library routine qsort, whose fourth argument is a routine that performs the comparison

```
qlcmp(q1,q2)
struct descrip **q1, **q2;
   {
   return (int)(StrLoc(**q1) - StrLoc(**q2));
   }
```

**Blocks.** After the location phase, some blocks in the allocated block region are marked and others are not. In the following typical situation, the horizontal lines delimit blocks, gray areas indicate marked blocks, and clear areas indicate unmarked blocks:

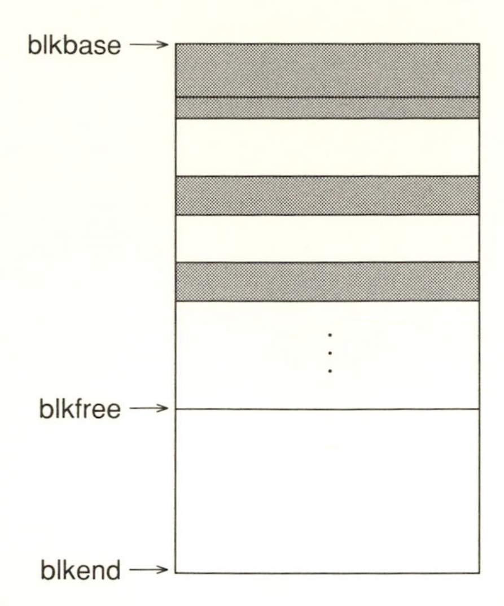


In the allocated block region, pointer adjustment and compaction are done in two linear passes over the region between blkbase and blkfree. In the first pass, two pointers are used, dest and source. dest points to where the next block will be after blocks are moved in the next pass, while source points to the next block to be processed. Both dest and source start at blkbase, pointing to the first allocated block.

During this pass, the title of each block pointed to by source is examined. If it is not marked (that is, if it is not larger than the maximum type code), dest is left unchanged and source is incremented by the size of the block to get to the title of the next block. Thus, unmarked blocks are skipped. The array bsizes is used, as before, to determine block sizes.

If the title of the block pointed to by source is marked, its back chain of descriptors is processed, changing their v-words to point to where dest points. In the case of a variable descriptor that is not a trapped-variable descriptor, the offset in its d-word is added to its v-word, so that it points to the appropriate relative position with respect to dest.

The last descriptor in the back chain is identified by the fact that its v-word contains a type code (a value smaller than any possible pointer to the allocated block region). This type code is restored to the title of the block before the v-word is changed to point to the destination. An m flag is set in the title to distinguish it as a marked block, since the former marking method no longer applies, but the compaction phase needs to determine which blocks are to be moved.

After the back chain has been processed, all descriptors that point to the block now point to where the block *will be* when it is moved during the compaction phase. The block itself is not moved at this time. This is illustrated by the example given previously, in which three descriptors point to a record block. After marking, the situation is

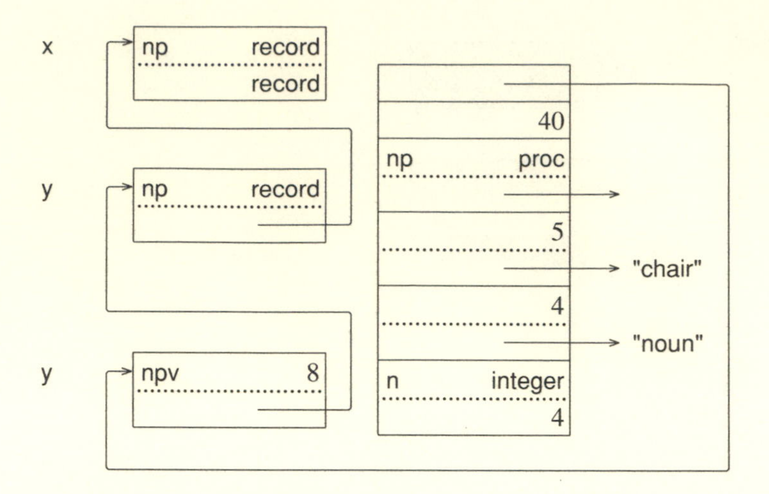


After processing the back chain, the situation is

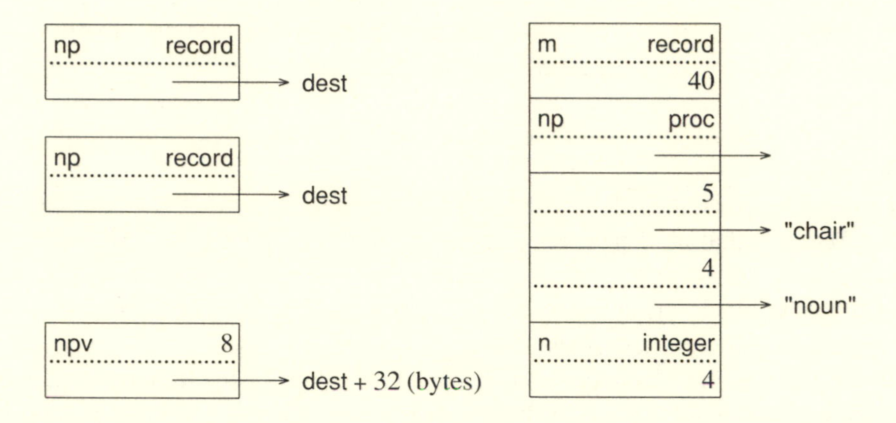


Note that the v-words of the descriptors point to where the block *will be* after it is moved.

The routine for adjusting pointers to the allocated block region is

```
adjust(source, dest)
char *source, *dest;
   {
   register struct descrip *nxtptr, *tptr;
```

```
/*
 * Loop through to the end of allocated block region, moving source
 * to each block in turn and using the size of a block to find the
 * next block.
 */
while (source < blkfree) {
   if ((uword)(nxtptr = (struct descrip *)BlkType(source)) > MaxType) {

      /*
       * The type field of source is a back pointer. Traverse the
       * chain of back pointers, changing each block location from
       * source to dest.
       */
      while ((uword)nxtptr > MaxType) {
         tptr = nxtptr;
         nxtptr = (struct descrip *)BlkLoc(*nxtptr);
         if (Var(*tptr) && !Tvar(*tptr))
            BlkLoc(*tptr) = (union block *)((word *)dest + Offset(*tptr));
         else
            BlkLoc(*tptr) = (union block *)dest;
         }
      BlkType(source) = (uword)nxtptr | F_Mark;
      dest += BlkSize(source);
      }
   source += BlkSize(source);
   }
}
```

When the pointer-adjustment phase is complete, the blocks can be moved. At this time, all the block titles contain type codes, and those that are to be saved are marked by m flags. During the compaction phase, these pointers are used again to reference the destination and source of blocks to be moved.

If an unmarked block is encountered, source is incremented by the block size, skipping over the block. If a marked block is encountered, the m flag in its title is removed and the block is copied to dest. Then dest and source are incremented by the size of the block.

When blkfree is reached, it is set to dest. At this point the allocated block region has been compacted. All saved blocks are before blkfree, and all free space is after it. The pointers that were adjusted now point to their blocks, and the relative situation is the same as it was before garbage collection.

The routine for compacting the allocated block region is

```
compact(source)
char *source;
   {
   register char *dest;
   register word size;

   /*
    * Start dest at source.
    */
   dest = source;

   /*
    * Loop through to end of allocated block space, moving source
    *  to each block in turn, using the size of a block to find the next
    *  block.  If a block has been marked, it is copied to the
    *  location pointed to by dest and dest is pointed past the end
    *  of the block, which is the location to place the next saved
    *  block.  Marks are removed from the saved blocks.
    */
   while (source < blkfree) {
      size = BlkSize(source);
      if (BlkType(source) & F_Mark) {
         BlkType(source) &= ~F_Mark;
         if (source != dest)
            mvc((uword)size, source, dest);
         dest += size;
         }
      source += size;
      }

   /*
    * dest is the location of the next free block.  Now that compaction
    *  is complete, point blkfree to that location.
    */
   blkfree = dest;
   }
```

The routine mvc(n, source, dest) moves n bytes from source to dest.

### 11.3.4 Collecting Co-Expression Blocks

After the location phase of garbage collection is complete, all the live co-expression blocks are marked, but the dead co-expression blocks are not. It is a simple matter to process the list of co-expression blocks, which are linked by pointers, calling free to deallocate dead ones and at the same time removing them from the list. For live co-expressions, the type code in the title is restored. The routine cofree that frees co-expression blocks is

```
cofree()
   {
   register struct b_coexpr **ep, *xep;
   extern int mstksize;                         /* main stack size */

   /*
    * Reset the type for &main.
    */
   BlkLoc(k_main)->coexpr.title = T_Coexpr;

   /*
    * The co-expression blocks are linked together through their
    *  nextstk fields, with stklist pointing to the head of the list.
    *  The list is traversed and each stack that was not marked
    *  is freed.
    */
   ep = &stklist;
   while (*ep != NULL) {
      if (BlkType(*ep) == T_Coexpr) {
         xep = *ep;
         *ep = (*ep)->nextstk;
         free((char *)xep);
         }
      else {
         BlkType(*ep) = T_Coexpr;
         ep = &(*ep)->nextstk;
         }
      }
   }
```

### 11.3.5 Expansion of the Allocated Regions

Garbage collection may not produce enough free space in a region to satisfy the request that caused the garbage collection. In this case, the region for which the request was made is expanded. In addition, the allocated string and block regions are expanded if the amount of free space in them after garbage collection otherwise would be less than a minimum value, which is called "breathing room." This expansion attempts to avoid "thrashing" that might result from a garbage collection that leaves a small amount of free space, only to result in a subsequent garbage collection almost immediately.

Since the allocated block region is at the end of the memory space for Icon, its expansion only involves calling the C library routine, sbrk, which expands the user's memory space. If this expansion fails, program execution is terminated with an error message.

If the allocated string region is expanded, however, the allocated block region must be relocated to make room. Relocating the block region requires relocating all pointers to it. No extra work is needed to do this, however. The relocation is accomplished by specifying the new location of the block region rather than the current blkbase as the second argument to adjust. Consequently, the adjusted pointers point to locations where blocks will be when they are moved at the end of garbage collection.

If the static region is expanded, both the allocated string region and the allocated block region must be relocated. The amount of the relocation for the allocated block region simply affects the second argument to adjust, as indicated previously. In the allocated string region, the amount is passed as the argument to scollect and is added to the v-words of the qualifiers pointed to from quallist, as indicated in Sec. 11.3.3.

### 11.3.6 Storage Requirements during Garbage Collection

Garbage collection itself takes some work space. Space for pointers to qualifiers is provided in quallist, while stack space is needed for calls to routines that perform the various aspects of garbage collection.

The space for quallist is obtained from the free space at the end of the allocated block region. The amount of space needed is proportional to the number of qualifiers whose v-words point to strings in the allocated string region and usually is comparatively small. Space for quallist is obtained in small increments. This is done in postqual, for which the complete routine is

```
postqual(dp)
struct descrip *dp;
   {
   extern char *brk();
   extern char *sbrk();
```

```
if (StrLoc(*dp) >= strbase && StrLoc(*dp) < strend) {
   /*
    * The string is in the string space.  Add it to the string qualifier
    *  list, but before adding it, expand the string qualifier list if
    *  necessary.
    */
   if (qualfree >= equallist) {
      equallist += Sqlinc;
      if ((int) brk(equallist) == -1)
         runerr(303, NULL);        /* terminate if expansion fails */
      currend = sbrk(0);
      }
   *qualfree++ = dp;
   }
}
```

The amount of stack space required during garbage collection depends primarily on the depth of recursion in calls to markblock; this is the only place in the garbage collection where recursion occurs. Recursion in markblock corresponds to linked lists of pointers in allocated storage. It occurs where a descriptor in the static region or the allocated block region points to an as-yet unmarked block. C stack overflow may occur during garbage collection. This problem is particularly serious on computers with small address spaces for programs that use a large amount of allocated data.

During garbage collection, the context for evaluation is switched to the C stack for &main, since allocated co-expression stacks are small, compared with the size of the stack for &main. The use of stack space during marking is minimized by testing descriptor v-words before calling markblock, by using static storage for variables in markblock that are not needed in recursive calls, and by incorporating the code for processing co-expression blocks in markblock, rather than calling a separate routine.

## 11.4 PREDICTIVE NEED

In most systems that manage allocated storage dynamically, garbage collections are triggered by allocation requests that cannot be satisfied by the amount of free storage that remains. In these systems, garbage collections occur during calls to allocation routines.

Whenever a garbage collection occurs, all potentially accessible data must be reachable from the basis, and any descriptors that are reachable from the basis must contain valid data. These requirements pose serious difficulties, since, in the normal course of computation, pointers to accessible objects may only exist in registers or on the C stack as C local variables that the garbage collector has no way of locating. Furthermore, descriptors that are being constructed may temporarily hold invalid data. While it is helpful to know that garbage collection can occur only during calls to allocation routines, allocation often is done in the midst of other computations. Assuring that all accessible data is reachable and that all reachable data is valid can be difficult and prone to error.

For these reasons, Icon uses a slightly different strategy, called "predictive need," for triggering garbage collections. Instead of garbage collection occurring as a byproduct of an allocation request, the amount of space needed is requested in advance. There are two routines, blkreq and strreq, for reserving space in advance. These routines check the block and string regions, respectively, to assure the amount of free space needed is actually available. If it is not, they call the garbage collector. For example, strreq is

```
strreq(n)
uword n;
   {
   strneed = n;                     /* save in case of collection */
   if (n > strend - strfree)
      collect();
   }
```

The amount of space requested is saved in the global variable strneed. Since the space requested may be allocated in pieces, this global variable is decremented when space is allocated:

```
char *alcstr(s, slen)
register char *s;
register word slen;
   {
   register char *d;
   char *ofree;

   /*
    * See if there is enough room in the string space.
    */
   if (strfree + slen > strend)
      syserr("string allocation botch");
   strneed -= slen;
```

```
/*
 * Copy the string into the string space, saving a pointer to its
 *  beginning.  Note that s may be null, in which case the space
 *  is still to be allocated but nothing is to be copied into it.
 */
ofree = d = strfree;
if (s) {
   while (slen-- > 0)
      *d++ = *s++;
   }

else
   d += slen;
strfree = d;
return ofree;
}
```

If a garbage collection occurs, the values of strneed and a similar variable for the allocated block region are checked to be sure that enough space is collected to satisfy any remaining allocation requests.

Since a predictive need request assures an adequate amount of space, no garbage collection can occur during the subsequent allocation request. The advantage of having a garbage collection occur during a predictive need request rather during an allocation request is that a safe time can be chosen for a possible garbage collection. The amount of space needed (or at least an upper bound on it) usually is known before the storage is actually needed, and when all valid data can be located from the basis.

The function repl provides an example:

```
FncDcl(repl, 2)
   {
   register char *sloc;
   register int cnt;
   long len;
   char sbuf[MaxCvtLen];
   extern char *alcstr();

   /*
    * Make sure that Arg1 is a string.
    */
   if (cvstr(&Arg1, sbuf) == CvtFail)
      runerr(103, &Arg1);
```

```
/*
 * Make sure that Arg2 is an integer.
 */
switch (cvint(&Arg2, &len)) {

   /*
    * Make sure count is not negative.
    */
   case T_Integer:
      if ((cnt = (int)len) >= 0)
         break;
      runerr(205, &Arg2);

   case T_Long:
      runerr(205, &Arg2);

   default:
      runerr(101, &Arg2);
   }

/*
 * Make sure the resulting string will not be too long.
 */
if ((len * StrLen(Arg1)) > MaxStrLen)
   runerr(205, NULL);

/*
 * Return an empty string if Arg2 is 0.
 */
if (cnt == 0)
   Arg0 = emptystr;

else {
   /*
    * Ensure enough space for the replicated string and allocate
    *  a copy of s.  Then allocate and copy s n - 1 times.
    */
   strreq(cnt * StrLen(Arg1));
   sloc = alcstr(StrLoc(Arg1), StrLen(Arg1));
   cnt--;
   while (cnt--)
      alcstr(StrLoc(Arg1), StrLen(Arg1));
```

```
        /*
         * Make Arg0 a descriptor for the replicated string.
         */
        StrLen(Arg0) = (int)len * StrLen(Arg1);
        StrLoc(Arg0) = sloc;
        }
    Return;
    }
```

A disadvantage of predictive need is that the maximum amount of storage needed must be determined and care must be taken to make predictive need requests prior to allocation. These problems do not occur in storage-management systems where garbage collection is implicit in allocation.

RETROSPECTIVE: Storage management is one of the major concerns in the implementation of a run-time system in which space is allocated dynamically and automatically. Although many programs never garbage collect at all, for those that do, the cost of garbage collection may be significant.

The requirements of storage management have a significant influence on the way that data is represented in Icon, particularly in blocks. Aspects of data representation that may appear arbitrary in the absence of considerations related to storage management have definite uses during garbage collection.

The garbage collector can only identify a pointer by virtue of the fact that it is contained in the v-word of a descriptor. Consequently, two words are required for all situations in which there may be a pointer to a live object, even if this pointer has no representation as a source-language data object. For example, pointers to list-element blocks are twice as large as they would need to be just to reference list-element blocks.

While it is possible to devise more economical methods of representing such data at the expense of complexity and loss of generality, any method of representing data for which space is allocated automatically has some overhead.

Garbage collection is most expensive when there are many live objects that must be saved. For programs in which allocated storage is used transiently and in which there are few live objects, garbage collection is fast.

## EXERCISES

**11.1** Since the first word of a block contains its type code, why is there also a type code in a descriptor that points to it?

**11.2** Give an example of an Icon expression that changes the contents of a block that is allocated statically in the run-time system.

**11.3** Give an example of an Icon expression that changes data in the icode region.

**11.4** Why not combine global and static identifiers in a single array of descriptors in the icode region?

**11.5** Why are the names of global identifiers needed?

**11.6** Why is there no array for the names of static identifiers?

**11.7** How long can a string be?

**11.8** How many elements can a single list-element block hold?

**11.9** List all the regions of memory in which the following Icon data objects can occur:

- strings
- descriptors
- co-expression blocks
- other blocks

**11.10** List all the source-language operations in Icon that may cause the allocation of storage.

**11.11** Give an example of an expression for which it cannot be determined from the expression itself whether or not it allocates storage.

**11.12** List the block types for which block size may vary from one block to another.

**11.13** List all the types of blocks that may occur in the allocated block region.

**11.14** List all the types of blocks that may occur outside of the allocated block region.

**11.15** Give an example of an Icon program in which the only access path to a live object during garbage collection is a variable that points to an element in a structure.

**11.16** Give an example of an Icon program that constructs a circular pointer chain.

**11.17** Explain how it can be assured that all blocks in the allocated block region are at addresses that are larger than the maximum type code.

**11.18** Aside from the possibility of looping in the location phase of garbage collection, what are the possible consequences of processing the descriptors in a block more than once?

**11.19** What would happen if there were more than one pointer on quallist to the *same* qualifier?

**11.20** Because of the way that the Icon run-time system is written, blocks that are not in the allocated block region do not contain pointers to allocated objects. Consequently, the descriptors in such blocks do not have to be processed during garbage collection.

- What does this imply about access to such blocks?
- What changes would have to be made to the garbage collector if such blocks could contain pointers to allocated objects?

**11.21** There is one exception to the statement in the preceding exercise that blocks that are not in the allocated data region do not contain pointers to allocated objects. Identify this exception and explain how it is handled during garbage collection.

**11.22** In the allocated string region, pointer adjustment and compaction are done in one pass, while two passes are used in the allocated block region. Why are pointer adjustment and compaction not done in a single pass over the allocated block region?

**11.23** What would be the effect of failing to remove the m flag from a block title during the compaction of the allocated block region?

**11.24** If garbage collection cannot produce enough free space in the region for which the collection was triggered, program execution is terminated even if there is extra space in another region. Describe how to modify the garbage collector to avoid this problem.

**11.25** Write a program that requires an arbitrarily large amount of space for quallist.

**11.26** Write a program that causes an arbitrary amount of recursion in markblock during garbage collection.

**11.27** Write a program that produces an arbitrarily large amount of data that must be saved by garbage collection, and observe the results.

**11.28** Devise a more sophisticated method of preventing thrashing in allocation and garbage collection than the fixed breathing-room method.

**11.29** There is no mechanism to reduce the size of an allocated region that may be expanded during one garbage collection, but which has an excessive amount of free space after another garbage collection. Describe how to implement such a mechanism.

**11.30** Suppose that a garbage collection could occur during a call of any C routine from any other C routine. How would this complicate the way C routines are written?

**11.31** What might happen if

- The amount of storage specified in a predictive need request were larger than the amount subsequently allocated?
- The amount of storage specified in a predictive need request were smaller than the amount subsequently allocated?

**11.32** When a list-element block is unlinked as the result of a pop, get, or pull, can the space it occupies always be reclaimed by a garbage collection? What are the general considerations in answering questions such as these?

**11.33** A variable that refers to a descriptor in a block points directly to the descriptor, with an offset in its d-word to the head of the block in which the descriptor resides. Could it be the other way around, with a variable pointing to the head of the block and an offset to the descriptor? If so, what are the advantages and disadvantages of the two methods?

**11.34** Why does sweep process an interpreter stack from its sp to its base, rather than the other way around?

**11.35** As mentioned in Sec. 11.3, all regions are collected, regardless of the region in which space is needed. Discuss the pros and cons of this approach.

**11.36** Evaluate the cost of using two-word descriptors for all pointers to blocks, even when these pointers do not correspond to source-language values (as, for example, in the links among list-element blocks).

**11.37** The need to garbage-collect blocks that are allocated during program execution significantly affects the structure and organization of such blocks. Suppose that garbage collection were never needed. How could the structure and organizations of blocks be revised to save space?

**11.38** Discuss the pros and cons of having different regions for allocating blocks of different types.

**11.39** Some expressions, such as

```
while write(read())
```

result in a substantial amount of ''storage throughput,'' even though no space really needs to be allocated. Explain why this effect cannot be avoided in general and discuss its impact on program performance.

**11.40** Physical memory is becoming less and less expensive, and more computer architectures and operating systems are providing larger user address spaces. Discuss how very large user address spaces might affect allocation and garbage-collection strategies.

CHAPTER 12

# Run-Time Support Operations

PERSPECTIVE: Several features of Icon's run-time system cannot be compartmentalized as neatly as storage management but present significant implementation problems nonetheless. These features include type checking and conversion, dereferencing and assignment, input and output, and diagnostic facilities.

## 12.1 TYPE CHECKING AND CONVERSION

Type checking is relatively straightforward in Icon. If only one type is of interest, a test of the d-word is sufficient, as in

```
if (Type(Arg1) != T_List)
   runerr(108, &Arg1);
```

It is necessary to test the entire d-word, since a qualifier may have a length that is the same as a type code. The d-word test takes care of this, because all descriptors that are not qualifiers have n flags.

If different actions are needed for different types, a separate test is required for qualifiers, since there is no type code for strings. Selection according to type generally has the form:

```
if (Qual(Arg1))                          /* string */
      .
      .
      .
else switch (Type(Arg1)) {
   case T_List:                          /* list */
      .
      .
      .
```

The real problems lie in type conversion, not type checking. At the source-language level, type conversion can occur explicitly, as a result of type-conversion functions, such as string(x), or it may be implicit. Implicit type conversion occurs frequently in many kinds of computations. For example, numeric data may be read from files in the form of strings, converted to numbers in arithmetic computations, and then converted to strings that are written out. Many operations support this implicit type conversion, and they rely on type-conversion routines.

There are four types among which mutual conversion is supported: strings, csets, integers, and real numbers. The details of type conversion are part of the Icon language definition (Griswold and Griswold 1983). For example, when a cset is converted to a string, the characters of the resulting string are in lexical order. Some conversions are conditional and may succeed or fail, depending on the value being converted. For example, a real number can be converted to an integer only if its value is in the range of a C long. The conversions are illustrated in the following diagram, where dashed lines indicate conversions that are conditional:

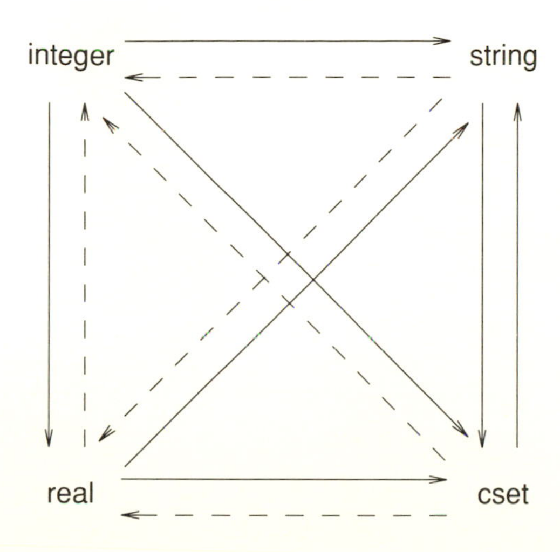


Thus, of the twelve conversions, five are conditional.

Some kinds of conversions are "natural" and occur frequently in typical programs. Examples are string-to-integer conversion and integer-to-string conversion. Other conversions, such as cset-to-integer, are unlikely to occur in the normal course of computation. To reduce the number of conversion routines required, these unlikely conversions are done in two steps. For example, the conversion of a cset to an integer is done by first converting the cset to a string and then converting the string to an integer. The direct conversions are

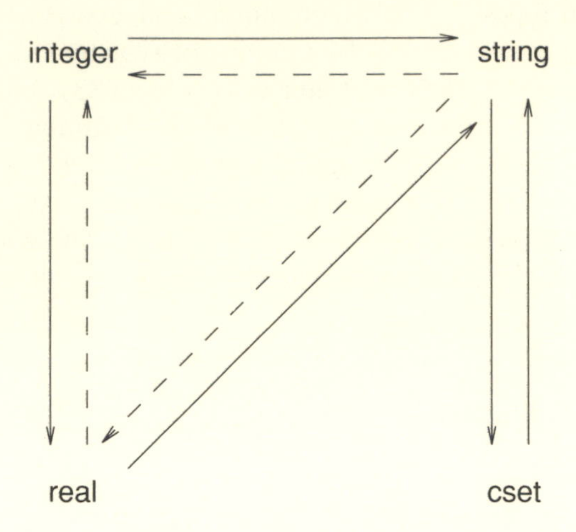


Conversions are done by calling routines that convert values to expected types. These routines are

| | |
|---|---|
| cvcset | convert to cset |
| cvint | convert to integer |
| cvreal | convert to real |
| cvstr | convert to string |

Since these routines may be called with any type of value, all of them are conditional. For example, it is not possible to convert a list to a string. These routines return the value NULL to indicate the failure of a conversion. If conversion is successful, they return a value indicating the type of conversion.

Numerical computation introduces complications in addition to the types integer and real, since there is the concept of a numeric "type" that includes both integers and real numbers. This is represented explicitly by the Icon type-conversion function numeric(x), which converts x to either integer or real, depending on the value of x. Numeric conversion occurs implicitly in polymorphic operations such as

```
n + m
```

which performs integer or real arithmetic, depending on the types of n and m. The routine for conversion of a value to numeric is

```
cvnum(dp, result)
register struct descrip *dp;
union numeric *result;
   {
   static char sbuf[MaxCvtLen];
```

```
    if (Qual(*dp)) {
        qtos(dp,  sbuf);
        return ston(sbuf, result);
        }

    switch (Type(*dp)) {

        case T_Integer:
            result->integer = (long)IntVal(*dp);
            return T_Integer;

        case T_Long:
            result->integer = BlkLoc(*dp)->longint.intval;
            return T_Long;

        case T_Real:
            GetReal(dp, result->real);
            return T_Real;

        default:
            /*
             * Try to convert the value to a string and
             *  then try to convert the string to an integer.
             */
            if (cvstr(dp, sbuf) == CvtFail)
                return CvtFail;
            return ston(StrLoc(*dp), result);
        }
    }
```

The macro GetReal is used to handle the real type, since some computers have restrictions on the alignment of doubles. Thus, GetReal has different definitions depending on the target computer. The actual conversion of a string to a numeric value is done by ston. Note that the conversion of a cset to a numeric value occurs in the default clause by way of conversion to a string.

The conversion routine cvstr requires a buffer to construct the string. This buffer is provided by the routine that calls cvstr, as illustrated by the previous example. See Sec. 4.4.4. The code for cvstr is

```
cvstr(dp, sbuf)
register struct descrip *dp;
char *sbuf;
    {
    double rres;
```

```
    if (Qual(*dp)) {
       return NoCvt;                    /* It is already a string */
       }

    switch (Type(*dp)) {
       /*
        * For types that can be converted into strings, call the
        *  appropriate conversion routine and return its result.
        *  Note that the conversion routines change the descriptor
        *  pointed to by dp.
        */
       case T_Integer:
          return itos((long)IntVal(*dp), dp, sbuf);

       case T_Long:
          return itos(BlkLoc(*dp)->longint.intval, dp, sbuf);

       case T_Real:
          GetReal(dp, rres);
          return rtos(rres, dp, sbuf);

       case T_Cset:
          return cstos(BlkLoc(*dp)->cset.bits, dp, sbuf);

       default:
          /*
           * The value cannot be converted to a string.
           */
          return CvtFail;
       }
    }
```

If dp is a qualifier, cvstr returns NoCvt, a code that indicates no conversion was performed. If a conversion is required, itos, rtos, or cstos does the actual work, placing its result in sbuf and changing the descriptor pointed to by dp accordingly. These routines return the code Cvt, which is, in turn, returned by cvstr. The value returned by cvstr therefore signals whether no conversion was needed (NoCvt), or a conversion was performed and the string is in sbuf (Cvt), or the conversion failed (NULL).

The reason that the return codes are needed is that if a converted string is in a buffer that is local to the calling routine, it must be copied into allocated storage. Otherwise it would be destroyed when that routine returns. A simple example of this situation occurs in the routine that implements the built-in function string(x):

```
FncDcl(string, 1)
   {
   char sbuf[MaxCvtLen];
   extern char *alcstr();

   Arg0 = Arg1;
   switch (cvstr(&Arg0, sbuf)) {

      /*
       * If Arg1 is not a string, allocate it and return it; if it is a
       *  string, just return it; fail otherwise.
       */
      case Cvt:
         strreq(StrLen(Arg0));              /* allocate converted string */
         StrLoc(Arg0) = alcstr(StrLoc(Arg0), StrLen(Arg0));

      case NoCvt:
         Return;

      default:
         Fail;
      }
   }
```

## 12.2 DEREFERENCING AND ASSIGNMENT

If there were no trapped variables, dereferencing and assignment would be trivial. For example, the descriptor d is dereferenced by

```
d = *VarLoc(d)
```

where VarLoc references the v-word of d:

```
#define VarLoc(d) ((d).vword.dptr)
```

The dereferencing or assignment to a trapped variable, on the other hand, may involve a complicated computation. This computation reflects the meaning associated with the operation on the source-language expression that is represented in the implementation as a trapped variable. For example, as discussed previously, in

```
x[y] := z
```

the value of x may be a list, a string, a table, or a record. A subscripted list or record does not produce a trapped variable, but the other two cases do. For a string, the variable on the left side of the assignment is a substring trapped variable. For a table, the variable is a table-element trapped variable. In the first case, the assignment involves the concatenation of three strings and the assignment of the result to x. In the second case, it involves looking for y in the table. If there is a table element for y, its assigned value is changed to the value of z. Otherwise, the table-element trapped-variable block is converted to a table-element block with the assigned value, and the block is inserted in the appropriate chain.

### 12.2.1 Dereferencing

Dereferencing of other trapped variables involves computations of comparable complexity. Dereferencing is done in the interpreter loop for arguments of operators for which variables are not needed. For example, in

```
n + m
```

the identifiers n and m are dereferenced before the function for addition is called (See Sec. 8.3.1). On the other hand, in

```
s[i]
```

the identifier i is dereferenced, but s is not, since the subscripting routine needs the variable as well as its value.

The function invocation routine also dereferences variables before a function is called. Note that there is no function that requires an argument that is a variable. Suspension and return from procedures also dereference local identifiers and arguments. Dereferencing occurs in a number of other places. For example, the function that handles subscripting must dereference the subscripted variable to determine what kind of result to produce.

The dereferencing routine begins as follows:

```
deref(dp)
struct descrip *dp;
   {
   register word i, j;
   register union block *bp;
   struct descrip v, tbl, tref;
   char sbuf[MaxCvtLen];
   extern char *alcstr();

   if (!Qual(*dp) && Var(*dp)) {
      /*
       * dp points to a variable and must be dereferenced.
       */
```

If dp does not point to a variable descriptor, the remaining code is skipped and deref simply returns.

If dp points to a variable that is not a trapped variable, dereferencing is simple:

```
if (!Tvar(*dp))
   /*
    * An ordinary variable is being dereferenced; just replace
    *  *dp with the descriptor *dp is pointing to.
    */
   *dp = *VarLoc(*dp);
```

There are three types of trapped variables with a switch on the type:

```
else switch (Type(*dp)) {

        case T_Tvsubs:
           /*
            * A substring trapped variable is being dereferenced.
            *  Point bp to the trapped variable block and v to
            *  the string.
            */
           bp = TvarLoc(*dp);
           v = bp->tvsubs.ssvar;
           DeRef(v);
           if (!Qual(v))
              runerr(103, &v);
           if (bp->tvsubs.sspos + bp->tvsubs.sslen - 1 > StrLen(v))
              runerr(205, NULL);
           /*
            * Make a descriptor for the substring by getting the
            *  length and pointing into the string.
            */
           StrLen(*dp) = bp->tvsubs.sslen;
           StrLoc(*dp) = StrLoc(v) + bp->tvsubs.sspos - 1;
           break;
```

The macro DeRef calls deref; a macro is used so that its definition can be changed to gather statistics on the use of dereferencing.

A table-element trapped variable may point to a table-element trapped-variable block or to a table-element block. The second situation occurs if two table-element trapped variables point to the same table-element trapped-variable block and assignment to one of the variables converts the table-element trapped-variable block to a table-element block before the second variable is processed. See Sec. 7.2. In this case, the value of the trapped variable is in the table-element block. On the other hand, if the trapped variable points to a table-element trapped-variable block, it is necessary to look up the subscripting value in the table, since an assignment for it may have been made between the time the trapped variable was created and the time it was dereferenced. If it is in the table, the corresponding assigned value is returned. If it is not in the table, the default assigned value is returned. The code is

```
case T_Tvtbl:
   if (BlkLoc(*dp)->tvtbl.title == T_Telem) {
      /*
       * The tvtbl has been converted to a telem and is
       *  in the table. Replace the descriptor pointed to
       *  by dp with the value of the element.
       */
      *dp = BlkLoc(*dp)->telem.tval;
      break;
      }

   /*
    * Point tbl to the table header block, tref to the
    * subscripting value, and bp to the appropriate
    * chain. Point dp to a descriptor for the default
    * value in case the value referenced by the subscript
    * is not in the table.
    */
   tbl = BlkLoc(*dp)->tvtbl.clink;
   tref = BlkLoc(*dp)->tvtbl.tref;
   i = BlkLoc(*dp)->tvtbl.hashnum;
   *dp = BlkLoc(tbl)->table.defvalue;
   bp = BlkLoc(BlkLoc(tbl)->table.buckets[SlotNum(i, TSlots)]);

   /*
    * Traverse the element chain looking for the subscript
    * value. If found, replace the descriptor pointed to
    * by dp with the value of the element.
    */
   while (bp != NULL && bp->telem.hashnum <= i) {
      if ((bp->telem.hashnum == i) &&
         (equiv(&bp->telem.tref, &tref))) {
            *dp = bp->telem.tval;
            break;
            }
      bp = BlkLoc(bp->telem.clink);
      }
   break;
```

The macro SlotNum(i, j) produces the slot number from the hash number, given j slots.

The last case, keyword trapped variables, is simpler, since the value is contained in the block pointed to by the trapped variable:

```
case T_Tvkywd:
   bp = TvarLoc(*dp);
   *dp = bp->tvkywd.kyval;
   break;
```

### 12.2.2 Assignment

The values of global identifiers are established initially as a byproduct of reading the icode file into the icode region. When procedures are called, the values of arguments and local identifiers are on the interpreter stack. These operations associate values with variables, but assignment, unlike dereferencing, is explicit in the source program.

The routine doasgn is used to perform all such operations. For example, the function for

```
x := y
```

is

```
OpDcl(asgn, 2, ":=")
   {
   /*
    * Make sure that Arg1 is a variable.
    */
   if (Qual(Arg1) || !Var(Arg1))
      runerr(111, &Arg1);

   /*
    * The returned result is the variable to which assignment is being
    *  made.
    */
   Arg0 = Arg1;

   /*
    * All the work is done by doasgn.  Note that Arg1 is known
    *  to be a variable.
    */
   if (!doasgn(&Arg1, &Arg2))
      Fail;
```

```
        Return;
        }
```

Note that assignment may fail. This can occur as the result of an out-of-range assignment to &pos and is indicated by a returned value of 0 from doasgn.

Like dereferencing, assignment is trivial for variables that are not trapped. The routine doasgn begins as follows:

```
doasgn(dp1, dp2)
struct descrip *dp1, *dp2;
   {
   register word i1, i2;
   register union block *bp;
   register struct b_table *tp;
   int (*putf)();
   union block *hook;
   char sbuf1[MaxCvtLen], sbuf2[MaxCvtLen];
   extern struct descrip tended[];
   extern struct b_lelem *alclstb();
   extern char *alcstr();

   tended[1] = *dp1;
   tended[2] = *dp2;
   ntended = 2;

assign:

   if (!Tvar(tended[1]))
      *VarLoc(tended[1]) = tended[2];
   else switch (Type(tended[1])) {
```

An array of descriptors in the basis is used, since garbage collection may occur at various places in doasgn, and it is important to assure that the descriptors pointed to by dp1 and dp2 are processed properly by the garbage collector. The value of the global variable ntended specifies how many tended descriptors are in use at any time.

As for dereferencing, there are three types of trapped variables to be considered. Assignment to a substring trapped variable is rather complicated:

```
case T_Tvsubs:
   /*
    * An assignment is being made to a substring trapped
    *  variable.  The tended descriptors are used as
    *  follows:
    *
    *    tended[1] - the substring trapped variable
    *    tended[2] - the value to assign
    *    tended[3] - the string containing the substring
    *    tended[4] - the substring
    *    tended[5] - the result string
    */

   /*
    * Be sure that the value to be assigned is a string.
    */
   ntended = 5;
   DeRef(tended[2]);
   if (cvstr(&tended[2], sbuf1) == CvtFail)
      runerr(103, &tended[2]);

   /*
    * Be sure that the variable in the trapped variable points
    *  to a string.
    */
   tended[3] = BlkLoc(tended[1])->tvsubs.ssvar;
   DeRef(tended[3]);
   if (!Qual(tended[3]))
      runerr(103, &tended[3]);
   strreq(StrLen(tended[3]) + StrLen(tended[2]));

   /*
    * Get a pointer to the substring trapped-variable block and
    *  make i1 a C-style index to the character that begins the
    *  substring.
    */
   bp = BlkLoc(tended[1]);
   i1 = bp->tvsubs.sspos - 1;

   /*
    * Make tended[4] a descriptor for the substring.
    */
   StrLen(tended[4]) = bp->tvsubs.sslen;
   StrLoc(tended[4]) = StrLoc(tended[3]) + i1;
```

```
/*
 * Make i2 a C-style index to the character after the
 *  substring. If i2 is greater than the length of the
 *  substring, it is an error because the string being
 *  assigned will not fit.
 */
i2 = i1 + StrLen(tended[4]);
if (i2 > StrLen(tended[3]))
   runerr(205, NULL);

/*
 * Form the result string.  First, copy the portion of the
 *  substring string to the left of the substring into the
 *  string space.
 */
StrLoc(tended[5]) = alcstr(StrLoc(tended[3]), i1);

/*
 * Copy the string to be assigned into the string space,
 *  effectively concatenating it.
 */
alcstr(StrLoc(tended[2]), StrLen(tended[2]));

/*
 * Copy the portion of the substring to the right of
 *  the substring into the string space, completing the
 *  result.
 */
alcstr(StrLoc(tended[3]) + i2, StrLen(tended[3]) - i2);

/*
 * Calculate the length of the new string.
 */
StrLen(tended[5]) = StrLen(tended[3]) - StrLen(tended[4]) +
   StrLen(tended[2]);
bp->tvsubs.sslen = StrLen(tended[2]);
tended[1] = bp->tvsubs.ssvar;
tended[2] = tended[5];

/*
 * Everything is set up for the actual assignment.  Go
 *  back to the beginning of the routine to do it.
 */
goto assign;
```

At the end of this case, no assignment has been made yet. Instead, tended[1] and tended[2] contain, respectively, the variable to which assignment is to be made and the value to be assigned. This is the same situation that exists at the beginning of the routine. The actual assignment is made by transferring back to the beginning. Note that, at this point, tended[1] may contain a trapped variable.

Table-element trapped variables have the same possibilities for assignment as for dereferencing. The processing is more complicated, since it may be necessary to convert a table-element trapped-variable block into a table-element block and link it into a chain:

```
case T_Tvtbl:
   /*
    *
    * The tended descriptors are used as follows:
    *
    *     tended[1] - the table element trapped variable
    *     tended[2] - the value to be assigned
    *     tended[3] - subscripting value
    */

   /*
    * Point bp to the trapped-variable block, point tended[3]
    *  to the subscripting value, and point tp to the table-
    *  header block.
    */
   ntended = 3;
   bp = BlkLoc(tended[1]);

   if (bp->tvtbl.title == T_Telem) {
      /*
       * The trapped-variable block already has been
       *  converted to a table-element block.  Just assign
       *  to it and return.
       */
      bp->telem.tval = tended[2];
      ntended = 0;
      return 1;
      }
   tended[3] = bp->tvtbl.tref;
   tp = (struct b_table *)BlkLoc(bp->tvtbl.clink);
```

```
/*
 * Get the hash number for the subscripting value and
 *  locate the chain that contains the element to which
 *  assignment is to be made.
 */
i1 = bp->tvtbl.hashnum;
i2 = SlotNum(i1, TSlots);
bp = BlkLoc(tp->buckets[i2]);

/*
 * Traverse the chain to see if the value is already in the
 *  table.  If it is there, assign to it and return.
 */
hook = bp;
while (bp != NULL && bp->telem.hashnum <= i1) {
  if (bp->telem.hashnum == i1 &&
     equiv(&bp->telem.tref, &tended[3])) {
        bp->telem.tval = tended[2];
        ntended = 0;
        return 1;
        }
   hook = bp;
   bp = BlkLoc(bp->telem.clink);
   }

/*
 * The value being assigned is new.  Increment the table
 *  size, convert the table-element trapped-variable block
 *  to a table-element block, and link it into the chain.
 */
tp->size++;
if (hook == bp) {             /* it goes at front of chain */
   bp = BlkLoc(tended[1]);
   bp->telem.clink = tp->buckets[i2];
   BlkLoc(tp->buckets[i2]) = bp;
   tp->buckets[i2].dword = D_Telem;
   }

else {                         /* it follows hook */
   bp = BlkLoc(tended[1]);
   bp->telem.clink = hook->telem.clink;
   BlkLoc(hook->telem.clink) = bp;
   hook->telem.clink.dword = D_Telem;
   }
```

```
        bp->tvtbl.title = T_Telem;
        bp->telem.tval = tended[2];
        ntended = 0;
        return 1;
```

In the case of a keyword trapped variable, the assignment routine that is pointed to from the keyword trapped-variable block is called to perform the assignment

```
    case T_Tvkywd:
        ntended = 2;
        putf = BlkLoc(tended[1])->tvkywd.putval;
        if ((*putf)(&tended[2]) == NULL) {
            ntended = 0;
            return 0;          /* assignment fails */
            }
        ntended = 0;
        return 1;
```

The assignment routine for &subject is typical:

```
putsub(dp)
struct descrip *dp;
   {
   char sbuf[MaxCvtLen];
   extern char *alcstr();

   switch (cvstr(dp, sbuf)) {

      case Cvt:
         strreq(StrLen(*dp));
         StrLoc(*dp) = alcstr(StrLoc(*dp), StrLen(*dp));

      case NoCvt:
         k_subject = *dp;
         k_pos = 1;
         break;

     default:
        runerr(103, dp);

      }

   return 1;
   }
```

## 12.3 INPUT AND OUTPUT

Icon supports only sequential file access. The run-time system uses C library routines to perform input and output, so the main implementation issues are those that relate to interfacing these routines.

### 12.3.1 Files

A value of type file in Icon points to a block that contains the usual title word, a FILE * reference to the file, a status word, and the string name of the file. The file status values are

| | |
|---|---|
| 0 | closed |
| 1 | open for reading |
| 2 | open for writing |
| 4 | open to create |
| 8 | open to append |
| 16 | open as a pipe |

These decimal numbers correspond to bits in the status word.

For example, the value of &input is

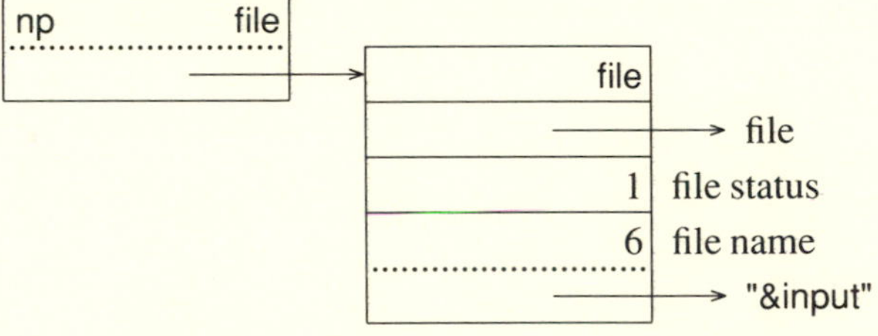


while the value of &output is

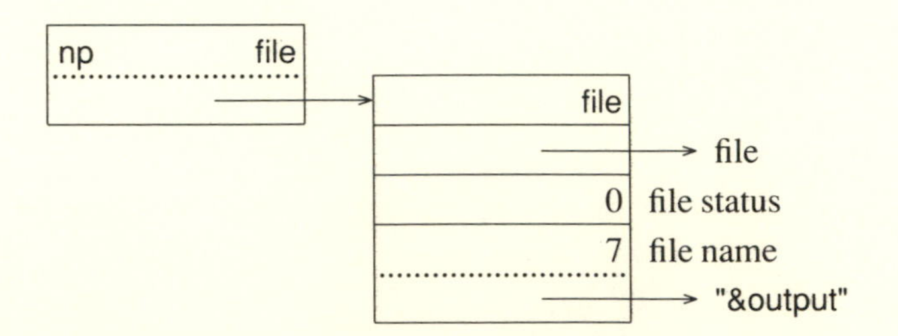


Another example is

out := open("log", "a")

for which the value of out is

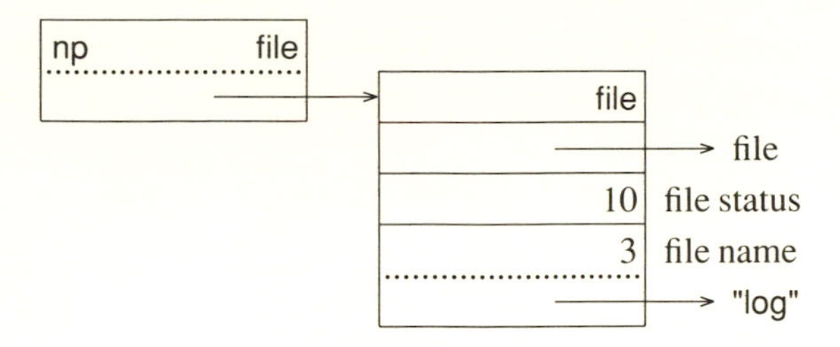


Note that the file status is 10, corresponding to being open for writing and appending.

Closing a file, as in

close(out)

merely changes its file status:

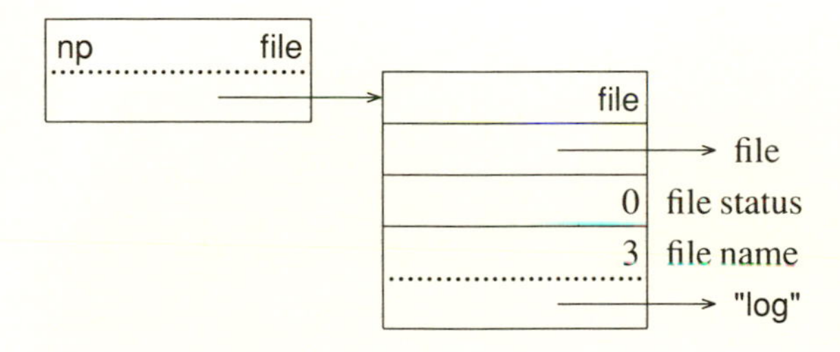


### 12.3.2 Reading and Writing Data

The function read(f) reads a line from the file f. In UNIX, a line is just a string of characters up to a newline character. There is no limit on the length of a line and the length of a line cannot be determined before it is read. On the other hand, there must be a place to store the line. Consequently, a limit is placed on the length of a line that can be read by read(f). This limit is an implementation parameter, which is usually 2,048 characters.

Characters are read into a buffer until a newline character is encountered or the limit is reached. A predictive need request is then made to assure that there is enough space in the allocated string region for the string, and the string is copied from the buffer into the string region.

The function reads(f, i) reads i characters from the file f. These characters may include newline characters. There is no limit on i; since the maximum length of the string to be read is known before characters are read. A predictive need request can be made to assure that there is enough space in the allocated string region. Characters are then read directly into the allocated string region without the use of an intervening buffer.

When strings are written, they are written directly from the allocated string region. There is no need to perform any allocation or to use an intermediate buffer.

Several strings can be concatenated on a file by

```
write(s1, s2, ..., sn)
```

This avoids the internal allocation and concatenation that is required for

```
write(s1 || s2 || ... || sn)
```

## 12.4 DIAGNOSTIC FACILITIES

Icon's diagnostic facilities consist of

- The function image(x), which produces a string representation of the value of x.
- The function display(f, i), which writes the names and values of identifiers in at most i levels of procedure call to the file f.
- Tracing of procedure calls, returns, resumptions, and suspensions.
- Run-time error termination messages.

Procedure tracing is done in invoke, pret, pfail, and psusp. If the value of &trace is nonzero, it is decremented and an appropriate trace message is written to standard error output. See Sec. 2.1.12 for an example.

The function display(f, i) must locate the names and values of local identifiers and arguments. The names are in the procedure block for the current procedure, which is pointed to by the zeroth argument of the current procedure call. The values are on the interpreter stack as described in Sec. 10.3.3.

Run-time termination messages are produced by the C routine runerr(n, dp), where dp is a pointer to the descriptor for the offending value. The value NULL is used for dp in cases where there is no offending value to print.

In all of these diagnostic situations, string representations of values are needed. The string representation for the ''scalar'' types string, cset, integer, and real is similar to what it is in the text of a source-language program. Long strings and csets are truncated to provide output that is easy to read. Other types present a variety of problems. For procedures, the type and procedure name are given.

A list, on the other hand, may be arbitrarily large and may contain values of any type, even lists. While the name may suffice for a procedure, often more information about a list is needed. As a compromise between information content and readability, only the first three and last three elements of a long list are included in its string representation. Since lists and other nonscalar types may be elements of lists, their representation as elements of a list is more restricted, with only the type and size being shown.

Since trace, display, and error output are written to files, the string representations can be written as they are determined, without regard for how long they are. The function image(x), on the other hand, returns a string value, and space must be allocated for it. A more limited form of string representation is used for nonscalar values, since the space needed might otherwise be very large.

## EXERCISES

**12.1** It is possible to conceive of meaningful ways to convert *any* type of data in Icon to any other. For example, a procedure might be converted to a string that consists of the procedure declaration. How would such a general conversion feature affect the way that types are converted in the run-time system?

**12.2** On computers with 16-bit words, Icon has two representations for integers internally (see Sec. 4.1.3). Describe how this complicates type conversion.

**12.3** How would the addition of a new numeric type, such as complex numbers, affect type conversion?

**12.4** How big would MaxCvtLen be if Icon had 512 different characters? 128? 64?

**12.5** Suppose a large-integer type were added to Icon to allow arithmetic on integers of an arbitrarily large size.

- How would this affect the numerical conversion routines?
- How would this affect cvstr and the routines that call it?

**12.6** List all the source-language operations that perform assignment.

**12.7** The routine doasgn returns the value 0 if assignment cannot be made. This signal results in a failure return in the functions that call doasgn (see Sec. 12.2.2). Why not use Fail in doasgn?

**12.8** Assuming that x, y, z, and w all have string values, diagram the structures that are produced in the course of evaluating the following expressions:

```
x[y] := z
z := x[y]
x[y] := z[w]
x[y][z] := w
```

Repeat this exercise for the case where all the identifiers have tables as values.

**12.9** Give an expression in which a table-element trapped variable points to a table-element block rather than to a table-element trapped-variable block.

**12.10** Give an expression in which a table-element trapped variable points to a table-element trapped-variable block, but where there is a table-element block in the table with the same entry value.

**12.11** Why are tended descriptors needed in doasgn but not in deref?

**12.12** Show an expression in which, at the end of the case for assignment to a substring trapped variable, the variable to which the assignment is to be made is a trapped variable. Can such a trapped variable be of any of the three types?

**12.13** Why is the string produced by read(f) not read directly into the allocated string region?

**12.14** Are there any circumstances in which write(x1, x2, ..., xn) requires the allocation of storage?

**12.15** Identify all the portions of blocks for source-language values that are necessary only for diagnostic output. How significant is the amount of space involved?

**12.16** The use of trapped variables for keywords that require special processing for assignment suggests that a similar technique might be used for substring and table-element trapped variables. Evaluate this possibility.

APPENDIX A

# Data Structures

This appendix summarizes, for reference purposes, all descriptor and block layouts in Icon.

## A.1 DESCRIPTORS

Descriptors consist of two words (normally C ints): a d-word and a v-word. The d-word contains flags in its most significant bits and small integers in its least significant bits. The v-word contains a value or a pointer. The flags are

| | |
|---|---|
| n | nonqualifier |
| p | v-word contains a pointer |
| v | variable |
| t | trapped variable |

### A.1.1 Values

There are three significantly different descriptor layouts for values. A qualifier for a string is distinguished from other descriptors by the lack of an n flag in its d-word, which contains only the length of the string. For example, a qualifier for the string "hello" is

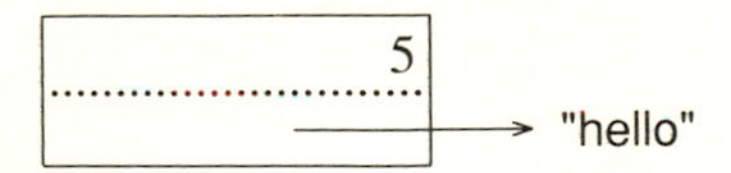


The null value and integers have type codes in their d-words and are self-contained. Examples are:

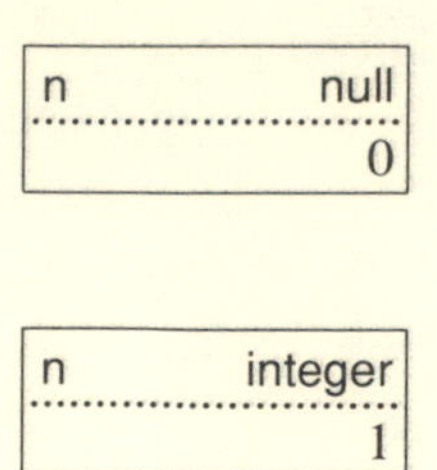


For all other data types, a descriptor contains a type code in its d-word and a pointer to a block of data in its v-word. A record is typical:

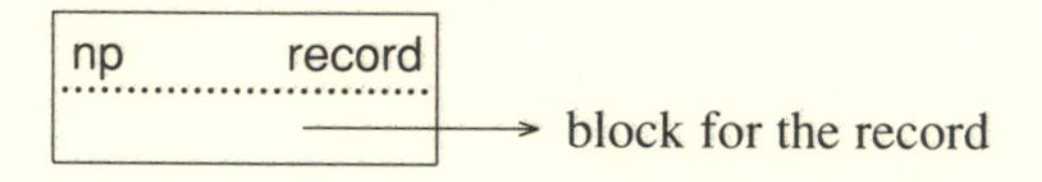


### A.1.2 Variables

There are two formats for variable descriptors. The v-word of an ordinary variable points to the descriptor for the corresponding value:

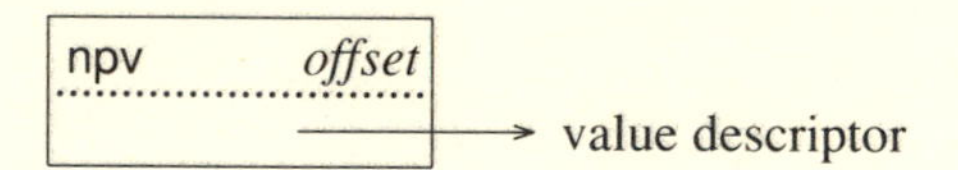


If the variable points to a descriptor in a block, the offset is the number of *words* from the top of the block to the value descriptor. If the variable points to a descriptor that corresponds to an identifier, the offset is zero.

The descriptor for a trapped variable contains a type code for the kind of trapped variable in its d-word and a pointer to the block for the trapped variable in its v-word. The trapped variable for &subject is typical:

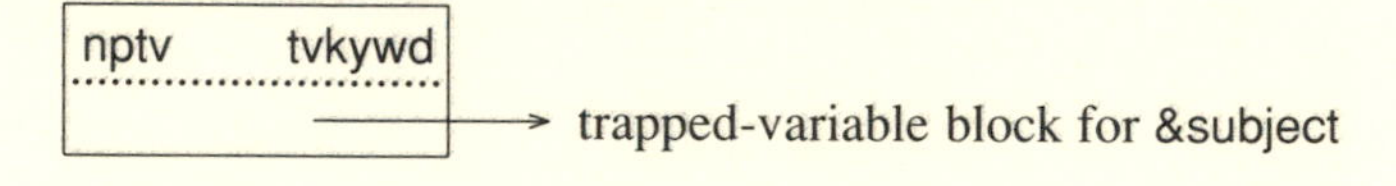

## A.2 BLOCKS

With the exception of the null value, integers, and strings, the data for Icon values is kept in blocks. The first word of every block is a title that contains the type code for the corresponding data type. For blocks that vary in size for a particular type, the next word is the size of the block in bytes. The remaining words depend on the block type, except that all non-descriptor data precedes all descriptor data. With the exception of the long integer block, the diagrams that follow correspond to blocks for computers with 32-bit words.

### A.2.1 Long Integers

On computers with 16-bit words, integers that are too large to fit in the d-word of a descriptor are stored in blocks. For example, the block for the integer 80,000 is

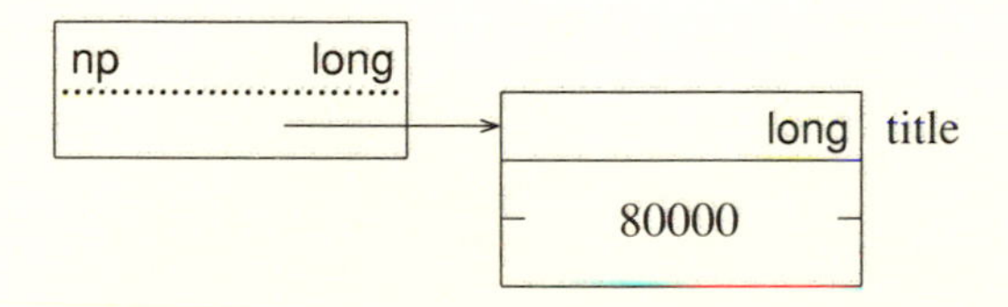


### A.2.2 Real Numbers

Real numbers are represented by C doubles. For example, on computers with 32-bit words, the real number 1.0 is represented by

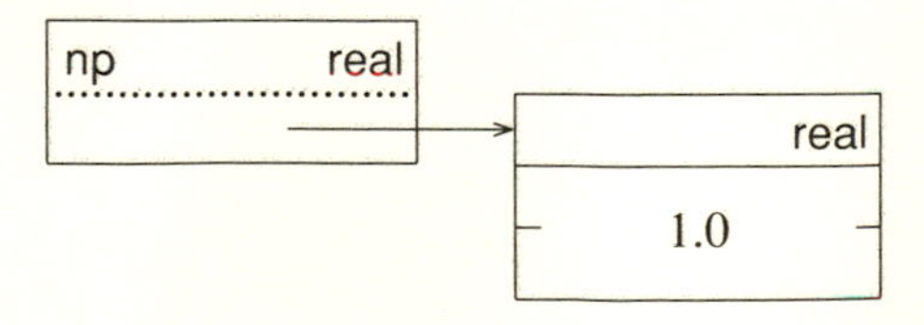

### A.2.3 Csets

The block for a cset contains the usual type code, followed by a word that contains the number of characters in the cset. Words totaling 256 bits follow, with a one in a bit position indicating that the corresponding character is in the cset, and a zero indicating that it is not. For example, &ascii is

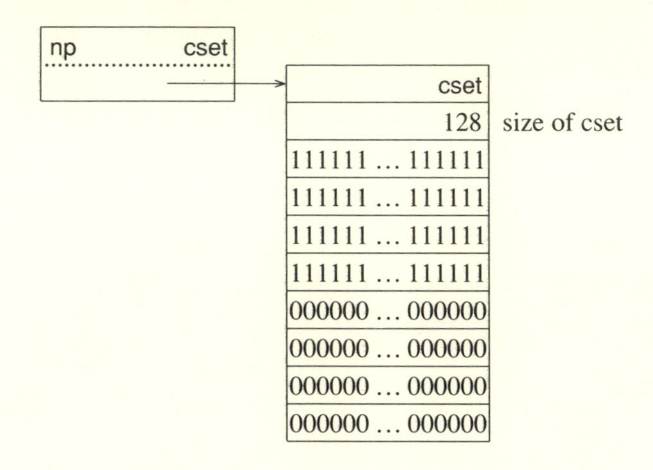


### A.2.4 Lists

A list consists of a list-header block that points to a doubly-linked list of list-element blocks, in which the list elements are stored in circular queues. See Chapter 6 for details. An example is the list

    [1,2,3]

which is represented as

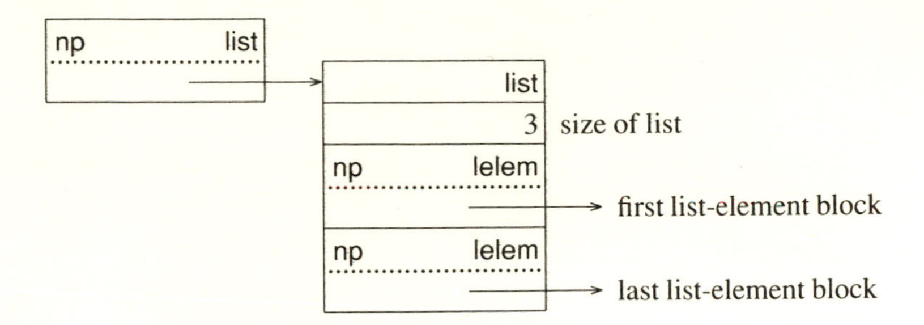


Here there is only one list-element block:

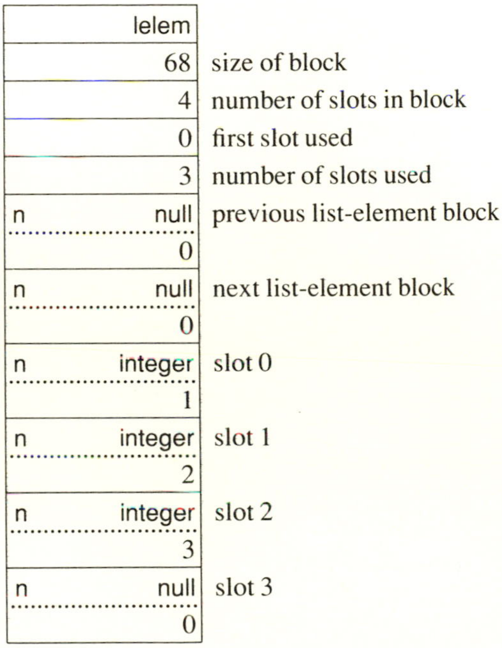

|  |  |
| --- | --- |
| lelem | |
| 68 | size of block |
| 4 | number of slots in block |
| 0 | first slot used |
| 3 | number of slots used |
| n null / 0 | previous list-element block |
| n null / 0 | next list-element block |
| n integer / 1 | slot 0 |
| n integer / 2 | slot 1 |
| n integer / 3 | slot 2 |
| n null / 0 | slot 3 |

### A.2.5 Sets

A set consists of a set-header block that contains slots for linked lists of set-element blocks. See Sec. 7.1 for details. An example is given by

```
set([1,2,3,4])
```

which is represented as

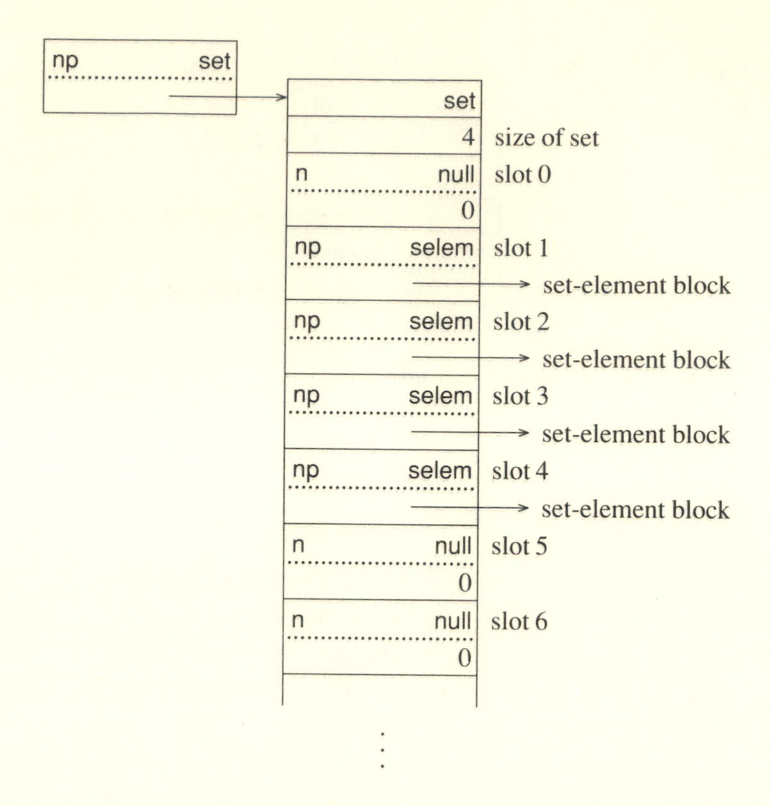


The set-element block for the member 3 is

| block | |
|---|---|
| selem | |
| 3 | hash number |
| n null<br>0 | next set-element block |
| n integer<br>3 | member value |

### A.2.6 Tables

A table is similar to a set, except that a table-header block contains the default assigned value as well as slots for linked lists of table-element blocks. See Sec. 7.2 for details. An example is given by

```
t := table()
every t[1 | 4 | 7] := 1
```

The table t is represented as

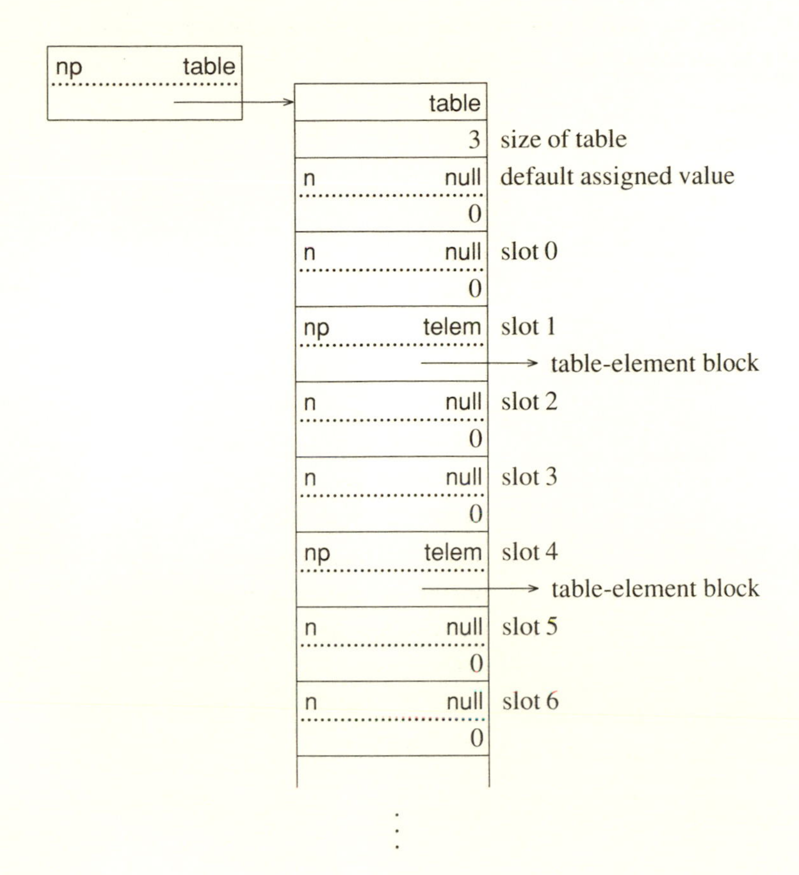


The table-element block for the entry value 4 in the previous example is

|  |  |
|---|---|
| telem | |
| 4 | hash number |
| n null | next table-element block |
| 0 | |
| n integer | entry value |
| 4 | |
| n integer | assigned value |
| 1 | |

### A.2.7 Procedures

The procedure blocks for procedures and functions are similar. For a procedure declaration such as

```
procedure calc(i, j)
   local k
   static base, index
        :
        :
end
```

the procedure block is

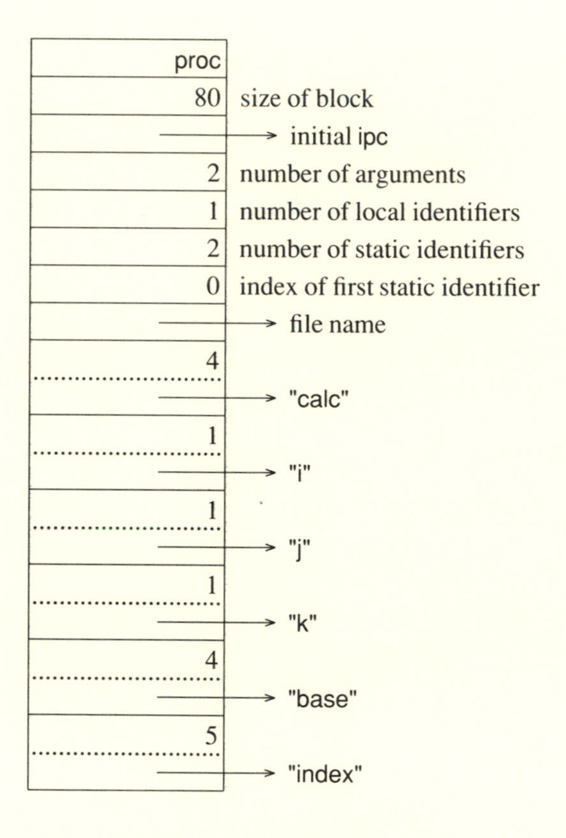


In a procedure block for a function, there is a value of −1 in place of the number of dynamic locals. For example, the procedure block for repl is

| | |
|---|---|
| proc | |
| 40 | size of block |
| → | C entry point |
| 2 | number of arguments |
| −1 | function indicator |
| 0 | not used |
| 0 | not used |
| 0 | not used |
| 4 | |
| → | "repl" |

In the case of a function, such as write, which has a variable number of arguments, the number of arguments is given as −1:

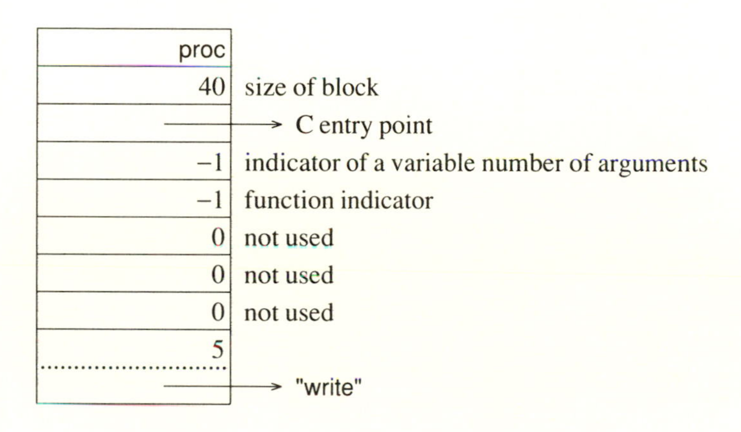


### A.2.8 Files

The block for a file contains a pointer to the corresponding file, a word containing the file status, and a qualifier for the name of the file. For example, the block for &output is

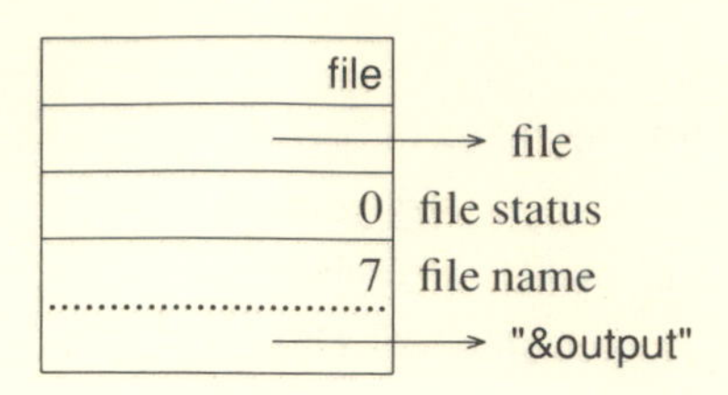


The file status values are

| | |
|---|---|
| 0 | closed |
| 1 | open for reading |
| 2 | open for writing |
| 4 | open to create |
| 8 | open to append |
| 16 | open as a pipe |

### A.2.9 Trapped Variables

There are three kinds of trapped variables: keyword trapped variables, substring trapped variables, and table-element trapped variables. The corresponding blocks are tailored to the kind of trapped variable.

The value of &trace illustrates a typical keyword trapped variable:

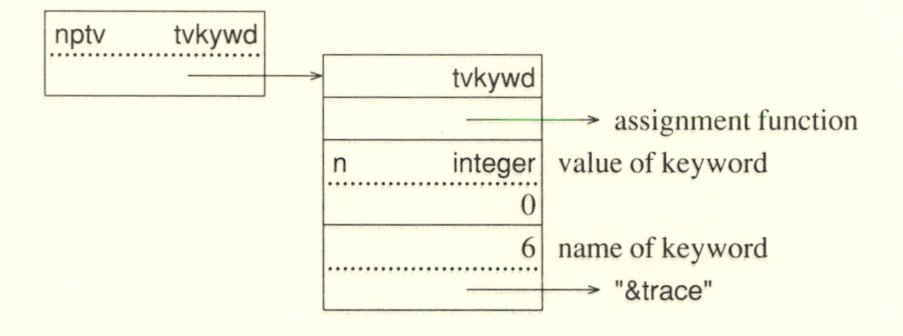


A substring trapped variable contains the offset and length of the substring, as well as a variable that points to the qualifier for the string. For example, if the value of s is "abcdef", the substring trapped-variable block for s[2:5] is

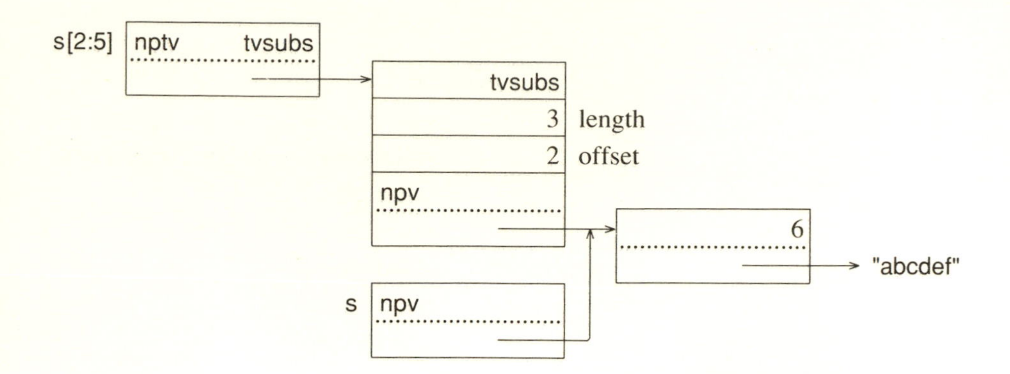


A table-element trapped-variable block contains a word for the hash number of the entry value, a pointer to the table, the entry value, and a descriptor reserved for the assigned value. For example, if t is a table, the table-element trapped-variable block for t[36] is

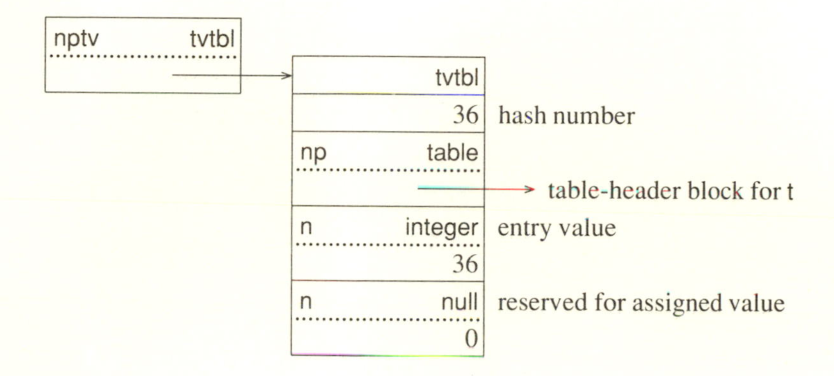


### A.2.10 Co-Expressions

A co-expression block consists of heading information, an array of words for saving the C state, an interpreter stack, and a C stack:

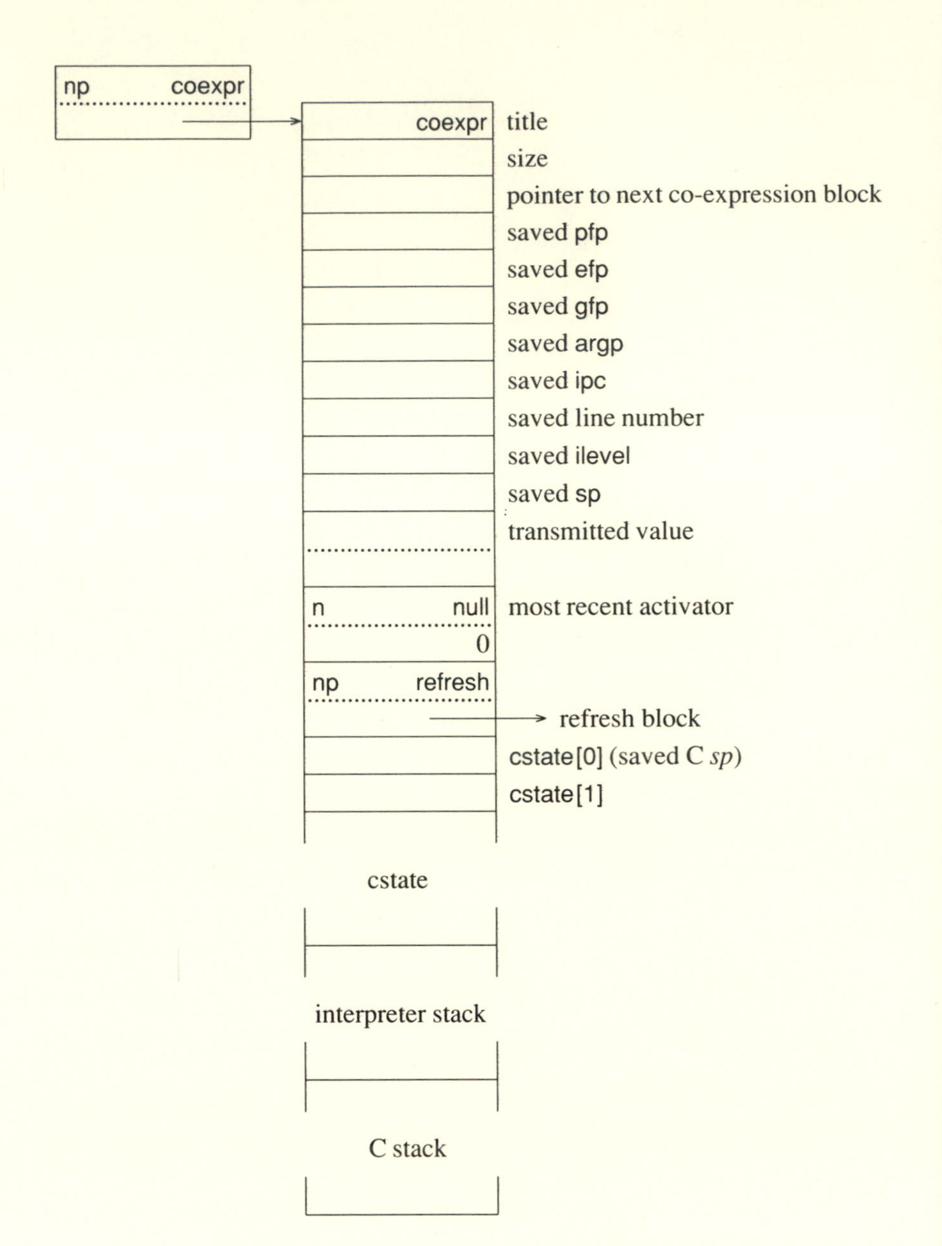


The refresh block contains information derived from the procedure block for the procedure in which the co-expression was created. Consider, for example,

```
procedure labgen(s)
   local i, j, e
   i := 1
   j := 100
   e := create (s || (i to j) || ":")
         :
         :
end
```

For the call labgen("L"), the refresh block for e is

|  |  |
|---|---|
| refresh | title |
| 88 | size of block |
| | initial ipc |
| 3 | number of local identifiers |
| 1 | number of arguments |
| | saved pfp |
| | saved efp |
| | saved gfp |
| | saved argp |
| | saved ipc |
| | saved line number |
| | saved ilevel |
| n proc | value of labgen |
| → | procedure block |
| 1 | value of s |
| → | "L" |
| n integer | value of i |
| 1 | |
| n integer | value of j |
| 100 | |
| n null | value of e |
| 0 | |

APPENDIX B

# Virtual Machine Instructions

This appendix lists all the Icon virtual machine instructions. For instructions that correspond to source-language operators, only the corresponding operations are shown. Unless otherwise specified, references to the stack mean the interpreter stack.

arg n — Push a variable descriptor pointing to argument n.

asgn — $expr_1$ := $expr_2$

bang — !*expr*

bscan — Push the current values of &subject and &pos. Convert the descriptor prior to these two descriptors into a string. If the conversion cannot be performed, terminate execution with an error message. Otherwise, assign it to &subject and assign 1 to &pos. Then suspend. If resumed, restore the former values of &subject and &pos and fail.

cat — $expr_1$ || $expr_2$

ccase — Push a copy of the descriptor just below the current expression frame.

chfail n — Change the failure ipc in the current expression frame marker to n.

coact — Save the current state information in the current co-expression block, restore state information from the co-expression block being activated, perform a context switch, and continue execution.

cofail — Save the current state information in the current co-expression block, restore state information from the co-expression block being activated, perform a context switch, and continue execution with co-expression failure signal set.

compl *~expr*

coret Save the current state information in the current co-expression block, restore state information from the co-expression block being activated, perform a context switch, and continue execution with co-expression return signal set.

create Allocate a co-expression block and a refresh block. Copy the current procedure frame marker, argument values, and local identifier values into the refresh block. Place a procedure frame for the current procedure on the stack of the new co-expression block.

cset a Push a descriptor for the cset block at address a onto the stack.

diff $expr_1$ -- $expr_2$

div $expr_1$ / $expr_2$

dup Push a null descriptor onto the stack and then push a copy of the descriptor that was previously on top of the stack.

efail If there is a generator frame in the current expression frame, resume its generator. Otherwise remove the current expression frame. If the ipc in its marker is nonzero, set ipc to it. If the failure ipc is zero, repeat efail.

eqv $expr_1$ === $expr_2$

eret Save the descriptor on the top of the stack. Unwind the C stack. Remove the current expression frame from the stack and push the saved descriptor.

escan Dereference the top descriptor on the stack if necessary. Copy it to the place on the stack prior to the saved values of &subject and &pos (see bscan). Exchange the current values of &subject and &pos with the saved values on the stack. Then suspend. If resumed, restore the values of &subject and &pos from the stack and fail.

esusp Create a generator frame containing a copy of the portion of the stack that is needed if the generator is resumed.

| | |
|---|---|
| field n | Replace the record descriptor on the top of the stack by a descriptor for field n of that record. |
| global n | Push a variable descriptor pointing to global identifier n. |
| goto n | Set ipc to n. |
| init n | Change init instruction to goto. |
| int n | Push a descriptor for the integer n. |
| inter | $expr_1$ ** $expr_2$ |
| invoke n | $expr_0(expr_1, expr_2, \ldots, expr_n)$ |
| keywd n | Push a descriptor for keyword n. |
| lconcat | $expr_1$ ||| $expr_2$ |
| lexeq | $expr_1$ == $expr_2$ |
| lexge | $expr_1$ >>= $expr_2$ |
| lexgt | $expr_1$ >> $expr_2$ |
| lexle | $expr_1$ <<= $expr_2$ |
| lexlt | $expr_1$ << $expr_2$ |
| lexne | $expr_1$ ~== $expr_2$ |
| limit | Convert the descriptor on the top of the stack to an integer. If the conversion cannot be performed or if the result is negative, terminate execution with an error message. If it is zero, fail. Otherwise, create an expression frame with a zero failure ipc. |
| line n | Set the current line number to n. |
| llist n | $[expr_1, expr_2, \ldots, expr_n]$ |
| local n | Push a variable descriptor pointing to local identifier n. |

lsusp — Decrement the current limitation counter, which is immediately prior to the current expression frame on the stack. If the limit counter is nonzero, create a generator frame containing a copy of the portion of the interpreter stack that is needed if the generator is resumed. If the limitation counter is zero, unwind the C stack and remove the current expression frame from the stack.

mark — Create an expression frame whose marker contains the failure ipc corresponding to the label n, the current efp, gfp, and ilevel.

mark0 — Create an expression frame with a zero failure ipc.

minus — $expr_1$ – $expr_2$

mod — $expr_1$ % $expr_2$

mult — $expr_1$ * $expr_2$

neg — –*expr*

neqv — $expr_1$ ~=== $expr_2$

nonnull — *expr*

null — /*expr*

number — +*expr*

numeq — $expr_1$ = $expr_2$

numge — $expr_1$ >= $expr_2$

numgt — $expr_1$ > $expr_2$

numle — $expr_1$ <= $expr_2$

numlt — $expr_1$ < $expr_2$

numne — $expr_1$ ~= $expr_2$

pfail — If &trace is nonzero, decrement it and produce a trace message. Unwind the C stack and remove the current procedure frame from the stack. Then fail.

plus $expr_1 + expr_2$

pnull Push a null descriptor.

pop Pop the top descriptor.

power $expr_1 \hat{} \; expr_2$

pret Dereference the descriptor on the top of the stack, if necessary, and copy it to the place where the descriptor for the procedure is. If &trace is nonzero, decrement it and produce a trace message. Unwind the C stack and remove the current procedure frame from the stack.

psusp Copy the descriptor on the top of the stack to the place where the descriptor for the procedure is, dereferencing it if necessary. Produce a trace message and decrement &trace if it is nonzero. Create a generator frame containing a copy of the portion of the stack that is needed if the procedure call is resumed.

push1 Push a descriptor for the integer 1.

pushn1 Push a descriptor for the integer −1.

quit Exit from the interpreter.

random $?expr$

rasgn $expr_1 <- expr_2$

real a Push a descriptor for the real number block at address a onto the stack.

refresh $\hat{}expr$

rswap $expr_1 <-> expr_2$

sdup Push a copy of the descriptor on the top of the stack.

sect $expr_1 [expr_2 : expr_3]$

size $*expr$

| | |
|---|---|
| static n | Push a variable descriptor pointing to static identifier n. |
| str n, a | Push a descriptor for the string of length n at address a. |
| subsc | $expr_1$[$expr_2$] |
| swap | $expr_1$ :=: $expr_2$ |
| tabmat | =*expr* |
| toby | $expr_1$ to $expr_2$ by $expr_3$ |
| unions | $expr_1$ ++ $expr_2$ |
| unmark | Remove the current expression frame from the stack and unwind the C stack. |
| value | .*expr* |

APPENDIX C

# Virtual Machine Code

The virtual machine code that is generated for various kinds of Icon expressions is listed below. The form of code given is icode, the output of the Icon linker, cast in a readable format. The ucode produced by the Icon translator, which serves as input to the Icon linker, is slightly different in some cases, since the linker performs some refinements.

## C.1 IDENTIFIERS

As mentioned in Sec. 8.2.2, the four kinds of identifiers are distinguished by where their values are located. All are referred to by indices, which are zero-based.

The values of global identifiers are kept in an array that is loaded from the icode file and is at a fixed place in memory during program execution. By convention, the zeroth global identifier contains the procedure descriptor for main. The following instruction pushes a variable pointing to the value of main onto the interpreter stack:

| *Icon expression* | *virtual machine code* |
|---|---|
| main | global 0 |

Static identifiers are essentially global identifiers that are only known on a per-procedure basis. Like global identifiers, the values of static identifiers are in an array that is at a fixed location. Static identifiers are numbered starting at zero and continuing through the program. For example, if count is static identifier 10, the following instruction pushes a variable descriptor pointing to that static identifier onto the stack:

| | |
|---|---|
| count | static 10 |

The space for the values of arguments and local identifiers is allocated on the stack when the procedure in which they occur is called. If x is argument zero and i is local zero for the current procedure, the following instructions push variable descriptors for them onto the stack:

| x | arg | 0 |
|---|---|---|

| i | local | 0 |
|---|---|---|

## C.2 LITERALS

The virtual machine instruction generated for an integer literal pushes the integer onto the stack as an Icon descriptor. The value of the integer is the argument to the instruction:

| 100 | int | 100 |
|---|---|---|

The instruction generated for a string literal is similar to that for an integer literal, except that the address of the string and its length are given as arguments. The string itself is in a region of data produced by the linker and is loaded as part of the icode file:

| "hello" | str | 5, a1 |
|---|---|---|

The instruction generated for a real or cset literal has an argument that is the address of a data block for the corresponding value. Such blocks are in the data region generated by the linker:

| 100.0 | real | a2 |
|---|---|---|

| 'aeiou' | cset | a3 |
|---|---|---|

## C.3 KEYWORDS

The instruction generated for most keywords results in a call to a C function that pushes a descriptor for the keyword onto the stack. The argument is an index that identifies the keyword. For example, &date is keyword 4:

| &date | keywd | 4 |
|---|---|---|

Some keywords correspond directly to virtual machine instructions. Examples are &null and &fail:

| &null | pnull |
|---|---|

| &fail | efail |
|---|---|

## C.4 OPERATORS

The code generated for a unary operator first pushes a null descriptor, then evaluates the code for the argument, and finally executes a virtual machine instruction that is specific to the operator:

| **expr* | pnull<br>*code for expr*<br>size |
|---|---|

The code generated for a binary operator is the same as the code generated for a unary operator, except that there are two arguments:

| $expr_1 + expr_2$ | pnull<br>*code for* $expr_1$<br>*code for* $expr_2$<br>plus |
|---|---|

An augmented assignment operator uses the virtual machine instruction dup to duplicate the result produced by its first argument:

| $expr_1$ +:= $expr_2$ | pnull<br>*code for* $expr_1$<br>dup<br>*code for* $expr_2$<br>plus<br>asgn |
|---|---|

The difference between the code generated for left- and right-associative operators is illustrated by the following examples:

| $expr_1 + expr_2 + expr_3$ | pnull<br>pnull<br>*code for* $expr_1$<br>*code for* $expr_2$<br>plus<br>*code for* $expr_3$<br>plus |
|---|---|

| $expr_1 := expr_2 := expr_3$ | pnull<br>*code for* $expr_1$<br>pnull<br>*code for* $expr_2$<br>*code for* $expr_3$<br>asgn<br>asgn |
|---|---|

A subscripting expression is simply a binary operator with a distinguished syntax:

| $expr_1 [expr_2]$ | pnull<br>*code for* $expr_1$<br>*code for* $expr_2$<br>subsc |
|---|---|

A sectioning expression is a ternary operator:

| $expr_1 [expr_2 : expr_3]$ | pnull<br>*code for* $expr_1$<br>*code for* $expr_2$<br>*code for* $expr_3$<br>sect |
|---|---|

Sectioning expressions with relative range specifications are simply abbreviations. The virtual machine instructions for them include the instructions for performing the necessary arithmetic:

| | |
|---|---|
| $expr_1$[$expr_2$+:$expr_3$] | pnull<br>*code for* $expr_1$<br>*code for* $expr_2$<br>dup<br>*code for* $expr_3$<br>plus<br>sect |

A to-by expression is another ternary operator with a distinguished syntax:

| | |
|---|---|
| $expr_1$ to $expr_2$ by $expr_3$ | pnull<br>*code for* $expr_1$<br>*code for* $expr_2$<br>*code for* $expr_3$<br>toby |

If the by clause is omitted, an instruction that pushes a descriptor for the integer 1 is supplied:

| | |
|---|---|
| $expr_1$ to $expr_2$ | pnull<br>*code for* $expr_1$<br>*code for* $expr_2$<br>push1<br>toby |

The code generated for an explicit list is similar to the code generated for an operator. The instruction that constructs the list has an argument that indicates the number of elements in the list:

| | |
|---|---|
| [$expr_1$, $expr_2$, $expr_3$] | pnull<br>*code for* $expr_1$<br>*code for* $expr_2$<br>*code for* $expr_3$<br>llist 3 |

## C.5 CALLS

The code generated for a call also is similar to the code generated for an operator, except that a null descriptor is not pushed (it is provided by the invoke instruction). The argument of invoke is the number of arguments present in the call, not counting the zeroth argument, whose value is the procedure or integer that is applied to the arguments:

| | |
|---|---|
| $expr_0(expr_1, expr_2)$ | *code for* $expr_0$<br>*code for* $expr_1$<br>*code for* $expr_2$<br>invoke 2 |

In a mutual evaluation expression in which the zeroth argument of the "call" is omitted, the default value is −1, for which an instruction is provided:

| | |
|---|---|
| $(expr_1, expr_2, expr_3)$ | pushn1<br>*code for* $expr_1$<br>*code for* $expr_2$<br>*code for* $expr_3$<br>invoke 3 |

## C.6 COMPOUND EXPRESSIONS AND CONJUNCTION

The difference between a compound expression and a conjunction expression is illustrated by the following examples. Note that the code generated for conjunction is considerably simpler than that generated for a compound expression, since no separate expression frames are needed:

| | |
|---|---|
| $\{expr_1; expr_2; expr_3\}$ | mark L1<br>*code for* $expr_1$<br>unmark<br>L1:<br>mark L2<br>*code for* $expr_2$<br>unmark<br>L2:<br>*code for* $expr_3$ |

| | |
|---|---|
| $expr_1$ & $expr_2$ & $expr_3$ | *code for* $expr_1$<br>pop<br>*code for* $expr_2$<br>pop<br>*code for* $expr_3$ |

## C.7 SELECTION EXPRESSIONS

In the code generated for an if-then-else expression, the control expression is bounded and has an expression frame of its own:

| | | | |
|---|---|---|---|
| if $expr_1$ then $expr_2$ else $expr_3$ | | mark | L1 |
| | | *code for* $expr_1$ | |
| | | unmark | |
| | | *code for* $expr_2$ | |
| | | goto | L2 |
| | L1: | | |
| | | *code for* $expr_3$ | |
| | L2: | | |

If the else clause is omitted, mark0 is used, so that if the control expression fails, this failure is transmitted to the enclosing expression frame:

| | |
|---|---|
| if $expr_1$ then $expr_2$ | mark0 |
| | *code for* $expr_1$ |
| | unmark |
| | *code for* $expr_2$ |

The code generated for a case expression is relatively complicated. As for similar control structures, the control expression is bounded. The result it produces is placed on the top of the stack by the eret instruction, which saves the result of evaluating $expr_1$, removes the current expression frame, and then pushes the saved result on the top of the stack. The ccase instruction pushes a null descriptor onto the stack and duplicates the descriptor just below the current efp on the top of the stack. This has the effect of providing a null descriptor and the first argument for the equivalence comparison operation performed by eqv. The second argument of eqv is provided by the code for the selector clause. The remainder of the code for a case clause removes the current expression frame marker, in case the comparison succeeds, and evaluates the selected expression:

| case $expr_1$ of {<br>  $expr_2$: $expr_3$<br>  $expr_4$: $expr_5$<br>  default: $expr_6$<br>  } |     mark0<br>    *code for* $expr_1$<br>    eret<br>    mark L2<br>    ccase<br>    *code for* $expr_2$<br>    eqv<br>    unmark<br>    pop<br>    *code for* $expr_3$<br>    goto L1<br>L2:<br>    mark L3<br>    ccase<br>    *code for* $expr_4$<br>    eqv<br>    unmark<br>    pop<br>    *code for* $expr_5$<br>    goto L1<br>L3:<br>    pop<br>    *code for* $expr_6$<br>L1: |
|---|---|

## C.8 NEGATION

The not control structure fails if its argument succeeds but produces the null value if its argument fails:

| not *expr* |     mark L1<br>    *code for expr*<br>    unmark<br>    efail<br>L1:<br>    pnull |
|---|---|

## C.9 GENERATIVE CONTROL STRUCTURES

If the first argument of an alternation expression produces a result, esusp produces a generator frame for possible resumption and duplicates the surrounding expression frame on the top of the stack. The result of the first argument is then pushed on the top of the stack, so that it looks as if the first argument merely produced a result. The second argument is then bypassed. When the first argument does not produce a result, its expression frame is removed, leaving the second argument to be evaluated:

| *expr₁ \| expr₂* | |
|---|---|
| | `        mark          L1`<br>`        `*code for expr₁*<br>`        esusp`<br>`        goto          L2`<br>`L1:`<br>`        `*code for expr₂*<br>`L2:` |

Since alternation is treated as a binary operation, a succession of alternations produces the following code:

| *expr₁ \| expr₂ \| expr₃* | |
|---|---|
| | `        mark          L1`<br>`        `*code for expr₁*<br>`        esusp`<br>`        goto          L2`<br>`L1:`<br>`        mark          L3`<br>`        `*code for expr₂*<br>`        esusp`<br>`        goto          L2`<br>`L3:`<br>`        `*code for expr₃*<br>`L2:` |

Repeated alternation is complicated by the special treatment of the case in which its argument does not produce a result. If it does not produce a result, the failure is transmitted to the enclosing expression frame, since the failure ipc is 0. However, if it produces a result, the failure ipc is changed by chfail so that subsequent failure causes transfer to the beginning of the repeated alternation expression. The esusp instruction produces the same effect as that for regular alternation. Note that changing the failure ipc only affects the expression frame marker on the stack. When mark is executed again, a new expression frame marker with a failure ipc of 0 is created.

| \|*expr* | L1:<br>mark0<br>*code for expr*<br>chfail L1<br>esusp |
|---|---|

In the limitation control structure, the normal left-to-right order of evaluation is reversed and the limiting expression is evaluated first. The limit instruction checks that the value is an integer and pushes it. It then creates an expression frame marker with a zero failure ipc. Thus, the limit is always one descriptor below the expression marker created by the subsequent mark instruction. The lsusp instruction is similar to the esusp instruction, except that it checks the limit. If the limit is zero, it fails instead of suspending. Otherwise, the limit is decremented:

| $expr_1$ \ $expr_2$ | *code for* $expr_2$<br>limit<br>*code for* $expr_1$<br>lsusp |
|---|---|

## C.10 LOOPS

The code generated for a repeat loop assures that the expression frame is handled uniformly, regardless of the success or failure of the expression:

| repeat *expr* | L1:<br>mark L1<br>*code for expr*<br>unmark<br>goto L1 |
|---|---|

A while loop, on the other hand, transmits failure to the enclosing expression frame if its control expression fails. Note that both $expr_1$ and $expr_2$ are evaluated in separate expression frames:

| | |
|---|---|
| while $expr_1$ do $expr_2$ | L1:<br>mark0<br>*code for* $expr_1$<br>unmark<br>mark L1<br>*code for* $expr_2$<br>unmark<br>goto L1 |

If the do clause is omitted, the generated code is similar to that for a repeat loop, except for the argument of mark:

| | |
|---|---|
| while *expr* | L1:<br>mark0<br>*code for expr*<br>unmark<br>goto L1 |

An until loop simply reverses the logic of a while loop:

| | |
|---|---|
| until $expr_1$ do $expr_2$ | L1:<br>mark L2<br>*code for* $expr_1$<br>unmark<br>efail<br>L2:<br>mark L1<br>*code for* $expr_2$<br>unmark<br>goto L1 |

The every-do control structure differs from the while-do control structure in that when its control expression produces a result, its expression frame is not removed. Instead, the result is discarded by pop, and the do clause is evaluated in its own expression frame. The efail instruction forces the resumption of a suspended generator that may have been produced by an esusp instruction in the code for $expr_1$:

| every $expr_1$ do $expr_2$ | mark0<br>*code for* $expr_1$<br>pop<br>mark0<br>*code for* $expr_2$<br>unmark<br>efail |
|---|---|

Breaks from loops normally occur in the context of other expressions. In the following example, the break expression removes the expression frame corresponding to the repeat control structure, evaluates its argument expression, and then transfers to a point beyond the end of the loop:

| repeat<br>$expr_1$ \| break $expr_2$ | L1:<br>mark L1<br>mark L3<br>*code for* $expr_1$<br>esusp<br>goto L4<br>L3:<br>unmark<br>*code for* $expr_2$<br>goto L2<br>L4:<br>unmark<br>goto L1<br>L2: |
|---|---|

Like break, next normally occurs in the context of other expressions. In the following example, next transfers control from a selection expression to the beginning of the loop:

| while $expr_1$ do<br>  if $expr_2$ then next<br>    else $expr_3$ | L1:<br>    mark0<br>    *code for* $expr_1$<br>    unmark<br>    mark L1<br>    mark L4<br>    *code for* $expr_2$<br>    unmark<br>    goto L2<br>L4:<br>    *code for* $expr_3$<br>L2:<br>    unmark<br>    goto L1 |
|---|---|

## C.11 STRING SCANNING

String scanning is a control structure, rather than an operator, since the values of &subject and &pos must be saved and new values established before the second argument is evaluated. This is accomplished by bscan. The instruction escan saves the current values of &subject and &pos and restores their values prior to the execution of bscan:

| $expr_1$ ? $expr_2$ | *code for* $expr_1$<br>bscan<br>*code for* $expr_2$<br>escan |
|---|---|

Augmented string scanning is similar to other augmented operations, but it differs in that the string scanning operation does not push a null value on the stack. The instruction sdup therefore is slightly different from dup, which is used in other augmented assignment operations:

| *expr₁* ?:= *expr₂* | |
|---|---|
| | pnull |
| | *code for expr₁* |
| | sdup |
| | bscan |
| | *code for expr₂* |
| | escan |
| | asgn |

## C.12 PROCEDURE RETURNS

The code generated for a return expression consists of the pret instruction. However, it allows for failure of the argument of return, which is equivalent to fail:

| return *expr* | | | |
|---|---|---|---|
| | | mark | L1 |
| | | *code for expr* | |
| | | pret | |
| | L1: | | |
| | | pfail | |

| fail | pfail |
|---|---|

The code generated for the suspend expression is analogous to the code generated for alternation, except that the result is returned from the current procedure. The efail instruction causes subsequent results to be produced if the call is resumed:

| suspend *expr* | |
|---|---|
| | mark0 |
| | *code for expr* |
| | psusp |
| | efail |

## C.13 CO-EXPRESSION CREATION

The first instruction in the code generated for a create expression is a transfer around the code that is executed when the resulting co-expression is activated. The create instruction constructs a descriptor that points to the co-expression whose code is at the label given in its argument and pushes this descriptor on the stack. When the co-expression is activated the first time, evaluation starts at the label stored in the co-expression. The result that is on the top of the stack is popped, since transmission of a result to the first activation of a co-expression is meaningless. If *expr* produces a result, coret returns that result to the activating co-expression. If *expr* fails, cofail signals failure to the activating co-expression:

| | | | |
|---|---|---|---|
| create *expr* | | goto | L3 |
| | L1: | | |
| | | pop | |
| | | mark | L2 |
| | | *code for expr* | |
| | | coret | |
| | | efail | |
| | L2: | | |
| | | cofail | |
| | | goto | L2 |
| | L3: | | |
| | | create | L1 |

APPENDIX D

# Adding Functions and Data Types

Icon is designed so that new functions and data types can be added with comparative ease. Such additions require changes only to the run-time system; the translator and linker are not affected.

This appendix provides some guidelines for modifying the Icon run-time system and lists useful macro definitions and support routines. It is designed to be read in conjunction with the source code for the implementation. The material included here only touches on the possibilities. There is no substitute for actually implementing new features and spending time studying the more intricate parts of the Icon system.

## D.1 FILE ORGANIZATION

The Icon system is organized in a hierarchy. Under UNIX, the Icon hierarchy is rooted at v6 and is usually located at /usr/icon/v6. For other operating systems, Icon may be named differently. The v6 directory has several subdirectories that contain source code, test programs, documents, and so forth. The source code is in v6/src. There are five subdirectories in src:

| | |
|---|---|
| h | common header files |
| icont | command processor |
| iconx | run-time system |
| link | linker |
| tran | translator |

The subdirectory h holds header files that are included by files in the other subdirectories. The file h/rt.h is particularly important, since it contains most of the definitions and declarations used in the run-time system.

The rest of the code related to the run-time system is in the subdirectory iconx. The first letters of files in this subdirectory indicate the nature of their contents. Files that begin with the letter f contain code for functions, while files that begin with o contain code for operators. Code related directly to the interpretive process is in files that begin with the letter i. ''Library'' routines for operations such as list construction that correspond to virtual machine instructions are in files that begin with the letter l. Finally, files that begin with the letter r hold run-time support routines.

Within each category, routines are grouped by functionality. For example, string construction functions such as map are in fstr.c, while storage allocation and garbage collection routines are in rmemmgt.c.

## D.2 ADDING FUNCTIONS

There are several conventions and rules of protocol that must be followed in writing a new function. The situations that arise most frequently are covered in the following sections. The existing functions in f files in iconx provide many examples to supplement the information given here.

### D.2.1 Function Declarations

A function begins with a call of the macro FncDcl(name, n), where name is the name of the function as it is called in a source-language program, and n is the number of arguments for the function. For example,

```
FncDcl(map, 3)
```

appears at the beginning of the function map. This macro declares the procedure block for the function and provides the beginning of the declaration of a C function for the code that follows. The value of n appears in the procedure block and is used to assure that the number of arguments on the interpreter stack when the function is called is the same as the number of arguments that the function expects. See Sec. 10.3.

An X is prepended to the name given to avoid a collision with the names of other C routines in the run-time system. Thus, the C function that implements map is named Xmap. Although the Icon function map has three arguments, the corresponding C function has only one: cargp, which is a pointer to an array of descriptors on the interpreter stack. For example, FncDcl(map, 3) generates

```
Xmap(cargp)
register struct descrip *cargp;
```

Other macros are provided for referencing the descriptors: Arg0 is the descriptor into which the result of a function is placed before it returns, Arg1 is the first descriptor argument in the call of the function, Arg2 is the second descriptor argument, and so on. These macros conceptually refer to the arguments in a source-language call of the function. It is never necessary (or desirable) to refer to cargp directly.

Note that the descriptor at Arg0 initially points to the procedure block for the function (see Sec. 10.1). It is fair to assume that Arg1, Arg2, ..., Argi, where i arguments are specified in the declaration, contain valid descriptors. Nothing can be assumed about the nature of these descriptors, other than that they represent valid source-language values. Similarly, a function must place a valid descriptor in Arg0 before returning, overwriting the procedure descriptor.

The macros described previously allow functions to be written without worrying about the details of the interpreter stack. It is not important to know how these macros are actually defined; it is best to think of them in terms of the higher-level concepts they embody.

### D.2.2 Returning from a Function

A function returns control to the interpreter by use of one of three macros, Return, Suspend, or Fail, depending on whether the function returns, suspends, or fails, respectively. Return and Fail return codes that the interpreter uses to differentiate between the two situations. Suspend returns control to the interpreter by calling it, as described in Sec. 9.3.

The use of Return is illustrated by the following trivial function that simply returns its argument:

```
FncDcl(idem, 1)
   {
   Arg0 = Arg1;
   Return;
   }
```

For example,

```
write(idem("hello"))
```

writes hello.

The use of Suspend and Fail is illustrated by the following function, which generates its first and second arguments in succession:

```
FncDcl(gen2, 2)
   {
   Arg0 = Arg1;
   Suspend;
   Arg0 = Arg2;
   Suspend;
   Fail;
   }
```

For example,

```
every write(gen2("hello", "there"))
```

writes

```
hello
there
```

As illustrated previously, Fail is used when there is not another result to produce. It is safe to assume that Arg0, Arg1, ... are intact when the function is resumed to produce another result.

Most functions have a fixed number of arguments. Only write, writes, and stop in the standard Icon repertoire can be called with an arbitrary number of arguments. For a function that can be called with an arbitrary number of arguments, an alternative declaration macro, FncDclV(name), is used. When this macro is used, the function is called with two arguments: the number of arguments in the call and a pointer to the corresponding array of descriptors. For example, FncDclV(write) generates

```
Xwrite(nargs, cargp)
int nargs;
register struct descrip cargp;
```

Within such a function, Arg0 refers to the return value as usual, but the arguments are referenced using the macro Arg(n). For example, a function that takes an arbitrary number of arguments and suspends with them as values in succession is

```
FncDclV(gen)
   {
   register int n;

   for (n = 1; n <= nargs; n++) {
      Arg0 = Arg(n);
      Suspend;
      }
   Fail;
   }
```

For example,

```
every write(gen("hello","there", "!"))
```

writes

```
hello
there
!
```

Note the use of Fail at the end of the function; the omission of Fail would be an error, since returning by flowing off the end of the function would not provide the return code that the interpreter expects.

### D.2.3 Type Checking and Conversion

Some functions need to perform different operations, depending on the types of their arguments. An example is type(x):

```
FncDcl(type, 1)
   {
   if (Qual(Arg1)) {
      StrLen(Arg0) = 6;
      StrLoc(Arg0) = "string";
      }
   else {
      switch (Type(Arg1)) {

         case T_Null:
            StrLen(Arg0) = 4;
            StrLoc(Arg0) = "null";
            break;

         case T_Integer:
         case T_Long:
            StrLen(Arg0) = 7;
            StrLoc(Arg0) = "integer";
            break;

         case T_Real:
            StrLen(Arg0) = 4;
            StrLoc(Arg0) = "real";
            break;
                  .
                  .
                  .
         }
      }
   Return;
   }
```

As indicated by this function, the d-word serves to differentiate between types, except for strings, which require a separate test.

For most functions, arguments must be of a specific type. As described in Sec. 12.1, type conversion routines are used for this purpose. For example, the function tab(i) requires that i be an integer. It begins as follows:

```
FncDcl(tab, 1)
   {
   register word i, j;
   word t, oldpos;
   long l1;

   /*
    * Arg1 must be an integer.
    */
   if (cvint(&Arg1, &l1) == CvtFail)
      runerr(101, &Arg1);
```

Note that cvint is called with the addresses of Arg1 and l1. If the conversion is successful, the resulting integer is assigned to l1. As indicated by this example, it is the responsibility of a function to terminate execution by calling runerr if a required conversion cannot be made.

The routine cvstr, which converts values to strings, requires a buffer, which is supplied by the routine that calls it. See Sec. 4.4.4. This buffer must be large enough to hold the longest string that can be produced by the conversion of any value. This size is given by the defined constant MaxCvtLen. For example, the function to reverse a string begins as follows:

```
FncDcl(reverse, 1)
   {
   register char c, *floc, *lloc;
   register word slen;
   char sbuf[MaxCvtLen];
   extern char *alcstr();

   /*
    * Make sure that Arg1 is a string.
    */
   if (cvstr(&Arg1, sbuf) == CvtFail)
      runerr(103, &Arg1);
```

The buffer is used only if a nonstring value is converted to a string. In this case, Arg1 is changed to a qualifier whose v-word points to the converted string in sbuf. This string does not necessarily begin at the beginning of sbuf. In any event, after a successful call to cvstr, the argument is an appropriate qualifier, regardless of whether a conversion actually was performed.

### D.2.4 Constructing New Descriptors

Some functions need to construct new descriptors to return in Arg0. Sometimes it is convenient to construct a descriptor by assignment to its d- and v-words. Various macros are provided to simplify these assignments. As given in the function type previously, StrLen and StrLoc can be used to construct a qualifier. For example, to return a qualifier for the string "integer", the following code suffices:

```
StrLen(Arg0) = 7;
StrLoc(Arg0) = "integer";
Return;
```

Here, the returned qualifier points to a statically allocated C string.

There also are macros and support routines for constructing certain kinds of descriptors. For example, the macro

```
Mkint(i, dp);
```

constructs an integer descriptor containing the integer i in the descriptor pointed to by dp. The definition of Mkint depends on the word size of the computer. On 32-bit computers, Mkint simply produces assignments to the d-word and v-word of descriptor pointed to by dp. On computers with 16-bit words, which have both T_Integer and T_Long forms of integers, Mkint produces a call to a support routine.

### D.2.5 Default Values

Many functions specify default values for null-valued arguments. There are support routines for providing default values. For example,

```
defstr(Arg3, sbuf, &q);
```

changes Arg3 to the string given by the qualifier q in case Arg3 is null-valued. If Arg3 is not null-valued, however, its value is converted to a string, if possible, by defstr. If this is not possible, defstr terminates execution with an error message.

### D.2.6 Storage Allocation

Functions that construct new data objects often need to allocate storage. Allocation is done in the allocated string region or the allocated block region, depending on the nature of the object. Support routines are provided to perform the actual allocation.

As mentioned in Sec. 11.4, predictive need requests *must* be made before storage is actually allocated. The functions strreq(i) and blkreq(i) request i bytes of storage in the allocated string and block regions, respectively.

Such a request generally should be made as soon as an upper bound on the amount of storage needed is known. It is not necessary to know the exact amount, but the amount requested must be at least as large as the amount that actually will be allocated. For example, the function reads(f, i) requests i bytes of string storage, although the string actually read may be shorter.

**String Allocation.** The function alcstr(s, i) copies i bytes starting at s into the allocated string region and returns a pointer to the beginning of the copy. For example, a function double(s) that produces the concatenation of s with itself is written as follows:

```
FncDcl(double, 1)
   {
   register int slen;
   char sbuf[MaxCvtLen];
   extern char *alcstr();

   if (cvstr(&Arg1, sbuf) == NULL)
      runerr(103, &Arg1);
   slen = StrLen(Arg1);
   strreq(2 * slen);
   StrLen(Arg0) = 2 * slen;
   StrLoc(Arg0) = alcstr(StrLoc(Arg1), slen);
   alcstr(StrLoc(Arg1), slen);
   Return;
   }
```

If the first argument of alcstr is NULL, instead of being a pointer to a string, the space is allocated and a pointer to the beginning of it is returned, but nothing is copied into the space. This allows a function to construct a string directly in the allocated string region.

If a string to be returned is in a buffer as a result of conversion from another type, care must be taken to copy this string into the allocated string region—otherwise the string in the buffer will be overwritten on subsequent calls. Copying such strings is illustrated by the function string(x) given in Sec. 12.1.

**Block Allocation.** The routine alcblk(i) allocates i bytes in the allocated block region and returns a pointer to the beginning of the block. The argument of alcblk must correspond to a whole number of words. There are run-time support routines for allocating various kinds of blocks. These routines, in turn, call alcblk. Such support routines generally fill in part of the block as well. For example, alccset(i) allocates a block for a cset, fills in the title and size words, and zeroes the bits for the cset:

```
struct b_cset *alccset(size)
int size;
   {
   register struct b_cset *blk;
   register i;
   extern union block *alcblk();

   blk = (struct b_cset *)alcblk((word)sizeof(struct b_cset), T_Cset);
   blk->size = size;

   /*
    * Zero the bit array.
    */
   for (i = 0; i < CsetSize; i++)
     blk->bits[i] = 0;
   return blk;
   }
```

See Sec. D.5.5 for a complete list of block-allocation functions.

### D.2.7 Storage Management Considerations

In addition to assuring that predictive need requests are made before storage is allocated, it is essential to assure that all descriptors contain valid data at any time a garbage collection may occur, that all descriptors are accessible to the garbage collector, and that all pointers to allocated data are in the v-words of descriptors.

Normally, all the descriptors that a function uses are on the interpreter stack and are referenced as Arg0, Arg1, ... . Such descriptors are processed by the garbage collector. Occasionally, additional descriptors are needed for intermediate computations. If such descriptors contain pointers in their v-words, it is *not* correct to declare local descriptors, as in

```
FncDcl(mesh, 2)
   {
   struct descrip d1, d2;
          :
```

The problem with this approach is that d1 and d2 are on the C stack and the garbage collector has no way of knowing about them.

However, since all descriptors on the interpreter stack are accessible to the garbage collector, intermediate computations can be performed on descriptors on the interpreter stack. Extra descriptors for this purpose can be provided by increasing the number of arguments specified for the function. Thus,

```
FncDcl(mesh, 4)
```

makes Arg3 and Arg4 available for intermediate computations. The initial values of Arg3 and Arg4 will be null because of argument adjustment performed by invoke unless mesh is called with extra arguments.

Garbage collection can occur only during a predictive need request. However, a predictive need request can occur between the time a function suspends and the time it is resumed to produce another result. Consequently, if a pointer is kept in a C variable in a loop that is producing results by suspending, the pointer may be invalid when the function is resumed. Instead, the pointer should be kept in the v-word of a descriptor that is accessible to the garbage collector.

### D.2.8 Error Termination

An Icon program may terminate abnormally for two reasons: as the result of a source-language programming error (such as an invalid type in a function call), or as a result of an error detected in the Icon system itself (such as a descriptor that should have been dereferenced but was not).

In case a source-language error is detected, execution is terminated by a call of the form

```
runerr(i, &d);
```

where i is an error message number and d is the descriptor for the offending value. If there is no specific offending value, the second argument is 0.

The array of error message numbers and corresponding messages is contained in iconx/imain.c. If there is no appropriate existing error message, a new one can be added, following the guidelines given in Appendix D of Griswold and Griswold 1983.

In theory, there should be no errors in the Icon system itself, but no large, complex software system is totally free of errors. Some situations are recognizable as being potential sources of problems in case data does not have the expected values. In such situations, especially during program development, it is advisable to insert calls of the function syserr, which terminates execution, indicating that an error was detected in the Icon system, and prints its argument as an indication of the nature of the error. It is traditional to use calls of the form

```
syserr("mesh: can't happen");
```

so that when, in fact, the ''impossible'' does happen, there is a reminder of human frailty. More informative messages are desirable, of course.

### D.2.9 Header Files

If a new function is added to an existing f file in iconx, the necessary header files normally will be included automatically. If a new function is placed in a new file, that file should begin with

```
#include "../h/rt.h"
```

This header file includes three other header files:

| | |
|---|---|
| ../h/config.h | general configuration information |
| ../h/cpuconf.h | definitions that depend on the computer word size |
| ../h/memsize.h | definitions that depend on the computer address space |

All of these files contain appropriate information for the local installation, and no changes in them should be needed.

In rare cases, it may be necessary to include other header files. For example, a function that deals directly with garbage collection might need to include iconx/gc.h.

### D.2.10 Installing a New Function

Both the linker and the run-time system must know the names of all functions. This information is provided in the header file h/fdefs.h.

In order to add a function, a line of the form

```
FncDef(name)
```

must be inserted in h/fdefs.h in proper alphabetical order.

Once this insertion is made, the Icon system must be recompiled to take into account the code for the new function. The steps involved in recompilation vary from system to system. Information concerning recompilation is available in system-specific installation documents.

## D.3 ADDING DATA TYPES

Adding a new data type is comparatively simple, although there are several places where changes need to be made. Failure to make all the required changes can produce mysterious bugs.

### D.3.1 Type Codes

At present, type codes range from 0 to 18. Every type must have a distinct type code and corresponding definitions. These additions are made in h/rt.h. First, a T_ definition is needed. For example, if a Boolean type is added, a definition such as

```
#define T_Boolean        19
```

is needed. The value of MaxType, which immediately follows the type code definitions, must be increased to 19 accordingly. Failure to set MaxType to the maximum type code may result in program malfunction during garbage collection. See Sec. 11.3.2.

Next a D_ definition is needed for the d-word of the new type. For a Boolean type, this definition might be

```
#define D_Boolean        (T_Boolean | F_Nqual)
```

All nonstring types have the F_Nqual flag and their T_ type code. Types whose v-words contain pointers also have the F_Ptr flag.

### D.3.2 Structures

A value of a Boolean type such as the one suggested previously can be stored in the d-word of its descriptor. However, most types contain pointers to blocks in their v-words. In this case, a declaration of a structure corresponding to the block must be added to h/rt.h. For example, a new rational number data type, with the type code T_Rational, might be represented by a block containing two descriptors, one for the numerator and one for the denominator. An appropriate structure declaration for such a block is

```
struct b_rational {
   int title;
   struct descrip numerator;
   struct descrip denominator;
   };
```

Since rational blocks are fixed in size, no size field is needed. However, a vector type with code T_Vector in which different vectors have different lengths needs a size field. The declaration for such a block might be

```
struct b_vector {
   int title;
   int blksize;
   struct descrip velems[1];
   };
```

As mentioned in Sec. 4.4.2, the size of one for the array of descriptors is needed to avoid problems with C compilers. In practice, this structure conceptually overlays the allocated block region, and the number of elements varies from block to block.

Any new structure declaration for a block must be added to the declaration union block in h/rt.h.

### D.3.3 Information Needed for Storage Management

All pointers to allocated data must be contained in the v-words of descriptors, since this is the only way the garbage collector can locate them. Furthermore, all non-descriptor data must precede any descriptors in a block. The amount of non-descriptor data, and hence the location of the first descriptor in a block, must be the same for all blocks of a given type.

As described in Sec. 11.3.2, the garbage collector uses the array bsizes to determine the size of a block and the array firstd to determine the offset of the first descriptor in the block. These arrays are in iconx/rmemmgt.c. When a new data type is added, appropriate entries must be made in these arrays. Failure to do so may result in serious bugs that occur only in programs that perform garbage collection, and the symptoms may be mysterious.

There is an entry in bsizes for each type code. If the type has no block, the entry is −1. If the type has a block of constant size, the entry is the size of the block. Otherwise, the entry is 0, indicating that the size is in the second word of the block. Thus, the entry for T_Boolean would be −1, the entry for T_Rational would be sizeof(struct b_rational), and the size for T_Vector would be 0.

There is a corresponding entry in firstd for each type code that gives the offset of the first descriptor in its corresponding block. If there is no block, the entry is −1. If the block contains no descriptors, the entry is 0. For example, the entry for T_Boolean would be −1, the entry for T_Rational would be WordSize, and the entry for T_Vector would be 2*WordSize, where WordSize is a defined constant that is the number of bytes in a word.

A third array, blknames, provides string names for all block types. These names are only used for debugging, and an entry should be made in blknames for each new data type.

### D.3.4 Changes to Existing Code

In addition to any functions that may be needed for operating on values of a new data type, there are several functions and operators that apply to all data types and which may, therefore, need to be changed for any new data type. These are

| | |
|---|---|
| *x | size of x (in iconx/omisc.c) |
| copy(x) | copy of x (in iconx/fmisc.c) |
| image(x) | string image of x (in iconx/fmisc.c) |
| type(x) | string name of type of x (in iconx/fmisc.c) |

There is not a concept of size for all data types. For example, a Boolean value presumably does not have a size, but the size of a vector presumably is the number of elements it contains. The size of a rational number is problematical. Modifications to *x are easy; see Sec. 4.4.4.

There must be some provision for copying any value. For structures, such as vectors, physical copies should be made so that they are treated consonantly with other Icon structures. For other data types, the "copy" consists of simply returning the value and not making a physically distinct copy. This should be done for data types, such as Boolean, for which there are only descriptors and no associated blocks. Whether or not a copy of a block for a rational value should be made is a more difficult decision and depends on how such values are treated conceptually, at the source-language level. It is, of course, easiest not to make a physical copy.

Some image must be provided for every value. This image should contain enough information to distinguish values of different types and, where possible, to provide some useful additional information about the specific value. The amount of detail that it is practical to provide in the image of a value is limited by the fact that the image is a string that must be placed in the allocated string region.

The type must be provided for all values and should consist of a simple string name. For example, if x is a Boolean value, type(x) should produce "boolean". The coding for type is trivial; see Sec. D.2.3.

There also are several run-time support routines that must be modified for any new type:

| | |
|---|---|
| outimage | image for tracing (in iconx/rmisc.c) |
| order | order for sorting (in iconx/rcomp.c) |
| anycmp | comparison for sorting (in iconx/rcomp.c) |
| equiv | equivalence comparison (in iconx/rcomp.c) |

The image produced for tracing purposes is similar to that produced by image and must be provided for all data types. However, outimage produces output and is not restricted to constructing a string in allocated storage. It therefore can be more elaborate and informative.

There must be some concept of sorting order for every Icon value. There are two aspects to sorting: the relative order of different data types and the ordering among values of the same type. The routine order produces an integer that corresponds to the order of the type. If the order of a type is important with respect to other types, this matter must be given some consideration. For example, a rational number probably belongs among the numeric types, which, in Icon, sort before structure types. On the other hand, it probably is not important whether vectors come before or after lists.

The routine anycmp compares two values; if they have the same order, as defined previously, anycmp determines which is the "smaller." For example, Boolean "false" might (or might not) come before "true," but some ordering between the two should be provided. On the other hand, order among vectors probably is not important (or well-defined), and they can be lumped with the other structures in anycmp, for which ordering is arbitrary. Sometimes ordering can be quite complicated; a correct ordering of rational numbers is nontrivial.

The routine equiv is used in situations, such as table subscripting and case expressions, to determine whether two values are equivalent in the Icon sense. Generally speaking, two structure values are considered to be equivalent if and only if they are identical. This comparison is included in equiv in a general way. For example, equiv need not be modified for vectors. Similarly, for data types that have no corresponding blocks, descriptor comparison suffices; equiv need not be modified for Boolean values either. However, determining the equivalence of numeric values, such as rational numbers, requires some thought.

## D.4 DEFINED CONSTANTS AND MACROS

Defined constants and macros are used heavily in Icon to parameterize its code for different operating systems and computer architectures and to provide simple, high-level constructions for commonly occurring code sequences that otherwise would be complex and obscure.

These defined constants and macros should be used consistently when making additions to Icon instead of using *ad hoc* constructions. This improves portability, readability, and consistency.

Learning the meanings and appropriate use of the existing defined constants and macro definitions requires investment of time and energy. Once learned, however, coding is faster, simpler, and less prone to error.

### D.4.1 Defined Constants

The following defined constants are used frequently in the run-time system. This list is by no means exhaustive; for specialized constants, see existing functions.

| | |
|---|---|
| CsetSize | number of words needed for 256 bits |
| LogHuge | one plus the maximum base-10 exponent of a C double |
| LogIntSize | base-2 logarithm of number of bits in a C int |
| MaxCvtLen | length of the longest possible string obtained by conversion |
| MaxLong | largest C long |
| MaxShort | largest C short |
| MaxStrLen | longest possible string |
| MinListSlots | minimum number of slots in a list-element block |
| MinLong | smallest C long |
| MinShort | smallest C short |
| WordSize | number of bytes in a word |

### D.4.2 Macros

The following macros are used frequently in the run-time system. See iconx/rt.h for the definitions, and see existing routines for examples of usages.

| | |
|---|---|
| Arg(n) | nth argument to function |
| ArgType(n) | type code of nth argument to function |
| ArgVal(n) | integer value of v-word of nth argument to function |
| BlkLoc(d) | pointer to block from v-word of d |
| BlkSize(cp) | size of block pointed to by cp |
| BlkType(cp) | type code of block pointed to by cp |
| ChkNull(d) | true if d is a null-valued descriptor |
| CsetOff(b) | offset in a word of cset bit b |
| CsetPtr(b, c) | address of word c containing cset bit b |
| DeRef(d) | dereference d |
| EqlDesc(d1, d2) | true if d1 and d2 are identical descriptors |
| GetReal(dp, r) | get real number into r from descriptor pointed to by dp |
| IntVal(d) | integer value of v-word of d |
| Max(i, j) | maximum of i and j |
| Min(i, j) | minimum of i and j |
| Mkint(i, dp) | make integer from i in descriptor pointed to by dp |
| Offset(d) | offset from d-word of variable descriptor d |

| | |
|---|---|
| Pointer(d) | true if v-word of d is a pointer |
| Qual(d) | true if d is a qualifier |
| Setb(b, c) | set bit b in cset c |
| SlotNum(i, j) | Slot for hash number i given j total slots |
| StrLen(q) | length of string referenced by q |
| StrLoc(q) | location of string referenced by q |
| Testb(b, c) | true if bit b in cset c is one |
| Tvar(d) | true if d is a trapped variable |
| TvarLoc(d) | pointer to trapped variable from v-word of d |
| Type(d) | type code in d-word of d |
| Var(d) | true if d is a variable descriptor |
| VarLoc(d) | pointer to value descriptor from v-word of d |
| Vsizeof(x) | size of structure x less variable array at end |
| Vwsizeof(x) | size of structure x in words less variable array at end |
| Wsizeof(x) | size of structure x in words |

## D.5 SUPPORT ROUTINES

There are many support routines for performing tasks that occur frequently in the Icon run-time system. Most of these routines are in files in iconx that begin with the letter r. The uses of many of these support routines have been illustrated earlier; what follows is a catalog for reference.

### D.5.1 Comparison

The following routines in iconx/rcomp.c perform comparisons:

anycmp(dp1, dp2) — Compare the descriptors pointed to by dp1 and dp2 as Icon values in sorting order, returning a value greater than 0, 0, or less than 0 depending on whether the descriptor pointed to by dp1 is respectively greater than, equal to, or less than the descriptor pointed to by dp2.

equiv(dp1, dp2) — Test for equivalence of descriptors pointed to by dp1 and dp2, returning 1 if equivalent and 0 otherwise.

lexcmp(dp1, dp2) Compare string qualifiers pointed to by dp1 and dp2, returning a value greater than 0, 0, or less than 0 depending on whether the string referenced by dp1 is respectively greater than, equal to, or less than the string referenced by dp2.

numcmp(dp1, dp2, dp3) Compare the descriptors pointed to by dp1 and dp2 as numbers, putting the converted value of the number referenced by dp2 in the descriptor pointed to by dp3 and returning 0, 1, or −1 depending on whether the number referenced by dp1 is respectively greater than, equal to, or less than the number referenced by dp2.

### D.5.2 Type Conversion

The following routines in iconx/rconv.c perform type conversions:

cvcset(dp, csp, sbuf) Convert the descriptor pointed to by dp to a cset and point csp to it, using sbuf as a conversion buffer if necessary.

cvint(dp, ip) Convert the descriptor pointed to by dp to an integer and store the value in the location pointed to by ip, returning the type, if the conversion can be performed, but NULL otherwise.

cvnum(dp, xp) Convert the descriptor pointed to by dp to a numeric value and place the result in the location pointed to xp, returning the type, if the conversion can be performed, but NULL otherwise.

cvpos(i1, i2) Convert i1 to a positive value with respect to the length i2, returning 0 if the conversion is not possible.

cvreal(dp, rp) Convert the descriptor pointed to by dp to a real number, storing the result in the location pointed to by rp and returning the type, if the conversion can be performed, but NULL otherwise.

cvstr(dp, sbuf) — Convert the descriptor pointed to by dp to a string, using sbuf as a buffer if necessary, returning Cvt if a conversion was performed, NoCvt if a conversion was unnecessary, or NULL if the conversion cannot be performed.

gcvt(n, i, sbuf) — Convert the number n to a string in sbuf, producing i significant digits if possible, otherwise using exponent notation.

mkreal(r, dp) — Make a real number descriptor for r in the descriptor pointed to by dp.

strprc(dp, i) — Convert the qualifier pointed to by dp to a procedure descriptor if possible, using i as the number of arguments in the case of a string that represents an operator, returning 0 if the conversion cannot be performed.

### D.5.3 Defaults

The following routines in iconx/default.c produce default values for omitted arguments:

defcset(dp, csp, sbuf, ip) — If the descriptor pointed to by dp is null, store the cset pointed to by ip at the place pointed to by csp and return 1; otherwise convert the descriptor pointed to by dp to a cset and return 0, but terminate with Error 104 if the conversion cannot be performed.

deffile(dp1, dp2) — If the descriptor pointed to by dp1 is null, replace it by the descriptor pointed to by dp2 and return 1; otherwise return 0 if dp1 points to a file descriptor but terminate with Error 105 otherwise.

defint(dp, ip, i) — If the descriptor pointed to by dp is null, store i at the location pointed to by ip and return 1. Otherwise convert the descriptor pointed to by dp to an integer, store it at the location pointed to by ip, and return 0, but terminate with Error 101 if the conversion cannot be performed.

defshort(dp, i) If the descriptor pointed to by dp is null, convert it to i and return 1. Otherwise convert the descriptor pointed to by dp to a short integer and return 0, but terminate with Error 101 if the conversion cannot be performed or Error 205 if the integer is too long.

defstr(dp1, sbuf, dp2) If the descriptor pointed to by dp1 is null, replace it by the descriptor pointed to by dp2 and return 0. Otherwise convert the descriptor pointed to by dp1 to a string and return 0, but terminate with Error 103 if the conversion cannot be performed.

### D.5.4 Assignment

The following routine in iconx/doasgn.c is used for all source-language assignments:

doasgn(dp1, dp2) Assign the descriptor pointed to by dp2 to the location referenced by the variable descriptor pointed to by dp1.

### D.5.5 Allocation

The following allocation routines in iconx/memmgt.c all return pointers to the objects they allocate:

alcblk(i, j) Allocate a block of i bytes with title j in the allocated block region.

alccoexp() Allocate a co-expression block.

alccset(i) Allocate a cset block for a cset, setting its size to i.

alcfile(fp, i, dp) Allocate a file block, setting its file pointer to fp, its status to i, and its name to the qualifier pointed to by dp.

alclint(i) Allocate a long-integer block and place i in it.

| | |
|---|---|
| alclist(i) | Allocate a list-header block and set its size field to i. |
| alclstb(i1, i2, i3) | Allocate a list-element block for i1 elements, setting its first field to i2 and its nused field to i3. |
| alcreal(r) | Allocate a real-number block and place r in it. |
| alcrecd(i, dp) | Allocate a record block with i fields, setting its procedure descriptor to the descriptor pointed to by dp. |
| alcrefresh(ip, i, j) | Allocate a refresh block for a procedure with i arguments, j local identifiers, and entry point ip. |
| alcselem(dp, i) | Allocate a set-element block, setting its member field to the descriptor pointed to by dp and its hash number field to i. |
| alcset() | Allocate a set-header block. |
| alcstr(sbuf, i) | Allocate a string of length i, and copy the string in sbuf into it, provided sbuf is not NULL. |
| alcsubs(i, j, dp) | Allocate a substring trapped-variable block, setting its length field to i, its offset field to j, and its variable descriptor to the descriptor pointed to by dp. |
| alctable(dp) | Allocate a table-header block, setting its default descriptor to the descriptor pointed to by dp. |
| alctelem() | Allocate a table-element block. |
| alctvtbl(dp1, dp2, i) | Allocate a table-element trapped-variable block, setting its link field to the descriptor pointed to by dp1, its entry field to the descriptor pointed to by dp2, and its hash number field to i. |
| blkreq(i) | Request i bytes of free space in the allocated block region. |

strreq(i) — Request i bytes of free space in the allocated string region.

### D.5.6 Operations on Structures

The following routines in iconx/rstruct.c perform operations on structures:

addmem(sp, ep, dp) — Add the set-element block pointed to by ep to the set pointed to by sp at the place pointed to by dp.

cplist(dp1, dp2, i, j) — Copy the sublist from i to j of the list referenced by the descriptor pointed to by dp1, and place the result in the descriptor pointed to by dp2.

locate(sp1, sp2) — Return 1 if the set-element block pointed to by sp2 is in the chain that starts at sp1, but return 0 otherwise.

memb(sp, dp, i, ip) — Set the value pointed to by ip to 1 if the descriptor pointed to by dp is a member of the set pointed by sp, using i as the hash number, but to 0 otherwise.

### D.5.7 Input and Output

The following routines in iconx/rsys.c perform input and output operations:

getstr(sbuf, i, fp) — Read a line of at most i characters from the file specified by fp, putting the result in sbuf, returning the number of characters read, but returning −1 on an end of file.

putstr(fp, sbuf, i) — Write i characters from sbuf on the file specified by fp.

### D.5.8 Error Termination

The following routines in iconx/imain.c cause error termination:

runerr(i, dp) — Terminate execution with error message i showing the offending value in the descriptor pointed to by dp, but omit it if dp is NULL.

syserr(sbuf) — Terminate execution with the system error message sbuf.

### D.5.9 Miscellaneous Operations

The following miscellaneous operations are in iconx/rmisc.c:

deref(dp) — Dereference the descriptor pointed to by dp.

hash(dp) — Return a hash value for the descriptor pointed to by dp.

outimage(fp, dp, i) — Write an image for the value of the descriptor pointed to by dp on the file pointed to by fp, but not calling outimage recursively if i is nonzero.

qtos(qp, sbuf) — Convert the string represented by the qualifier pointed to by qp to a null-terminated C-style string in sbuf.

### D.5.10 Diagnostics

There are two routines in iconx/rmemmgt.c for producing diagnostic output:

descr(dp) — Write a diagnostic diagram of the descriptor pointed to by dp to standard error output.

blkdump() — Write a diagnostic diagram of the allocated block region to standard error output.

## D.6 DEBUGGING

Debugging a modification to Icon can be difficult unless the overall structure of the Icon system is understood. It is especially important to understand the way Icon's data is organized and how storage management works. If an addition to Icon does not work properly, or if Icon itself no longer functions correctly after a modification, it generally is advisable to *think* about the possible sources of problems, instead of immediately resorting to a debugger.

Print statements are a crude but often effective means of locating the source of an error. When adding diagnostic output, use

```
fprintf(stderr, " ... ", ... );
```

instead of the corresponding printf. On some systems it may be useful to follow such calls by fflush(stderr) to assure that diagnostic output appears as soon as it is produced.

Icon normally traps floating-point exceptions and illegal addressing (segmentation violations), since these errors can result from source-language programming errors, such as division by real zero and excessive recursion resulting in C stack overflow. For example, Icon normally produces run-time Error 302 ("C stack overflow") in case of a segmentation violation. This method of handling such an error is appropriate if Icon itself is free of bugs, but it interferes with debugging in situations where there are likely to be bugs in new code.

Assigning a value to the environment variable ICONCORE turns off the trapping of such errors. In this case, most systems produce meaningful diagnostic messages and a core dump in the case of an exception. If ICONCORE has a value, a core dump also is produced if an Icon program terminates as a result of runerr or syserr. It therefore is good practice to set ICONCORE when making modifications to Icon. For an extended debugging session, it may be convenient to set dodump in iconx/imain.c to 1.

APPENDIX E

# Projects

Icon lends itself to extensions, both at the language level and in its implementation. Its numerous data types suggest additional types and ways of combining and relating them. Many changes can be made by modifying only the run-time system. All in all, Icon provides an unusually powerful tool for experimentation.

This appendix contains a collection of implementation projects. These range from writing comparatively simple functions to creating features that require extensive and pervasive changes to the implementation. Most of the projects involve only the run-time system. Some projects are research topics that have resisted previous efforts at solution. Part of the challenge in the projects that follow is to recognize the difficulties in and the relationship between design and implementation issues.

Some of the projects are grouped in general subject areas, such as numerical extensions. Others are larger in scope and stand alone. There is no particular unifying theme; in fact, some projects are incompatible with others.

The projects given here are only a sampling. No attempt has been made to cover all possible extensions and modifications to Icon. Instead, the projects here are intended to be both representative and suggestive.

## E.1 NUMERICAL EXTENSIONS

Icon emphasizes nonnumerical computation and supports only the conventional kinds of numerical computations that are found in most programming languages. Many problems, however, involve other kinds of numerical computation. For example, complex arithmetic is needed for the solutions of many problems, although only a few programming languages support it directly. There also are many problems in number theory that involve both arithmetic on integers of arbitrarily large size and at the same time involve pattern matching and the rearrangement of the digits of such integers. Considerations like these motivate the following new data types for Icon:

- complex numbers
- rational numbers
- integers of arbitrarily large size (''multiple-precision integers'')

Provide appropriate operations on values of each of these data types. Include type conversions in the spirit of the regular numeric types in Icon.

*Comments:* Data representations for complex and rational numbers are relatively straightforward. Complex arithmetic is relatively routine as long as it is based on integer arithmetic (as opposed, say, to complex arithmetic on rational numbers). Rational arithmetic quickly gets into practical difficulties, however (Knuth 1969, pp. 290-292). There are many possible data representations for large integers (Knuth 1969, pp. 229-280; Griswold and Griswold 1983, pp. 192-200); the choice of one involves many interacting considerations and should not be made hastily.

## E.2 STRING-PROCESSING EXTENSIONS

Although Icon has an extensive repertoire of operations on strings, there are many possible variations on string manipulation as well as some unusual extensions to the basic concept of a string.

**1.** Add the following string-processing functions:

| | |
|---|---|
| compress(s, c) | compress consecutive occurrences of characters in the cset c that occur in the string s |
| delete(s, c) | delete all occurrences of characters in the cset c that occur in the string s |
| rotate(s, i) | rotate the string s to the left i characters |
| field(s, c) | generate substrings of s that are separated by characters in the cset c |

Provide reasonable defaults for i in rotate and c in field.

**2.** Add a character data type. Provide automatic conversions so that this type is ''transparent'' and can be used in all operations on strings.

**3.** Add a bit-string data type and provide appropriate operations on values of this type. Consider what would be involved in performing bit-wise operations on integers and how type conversion should be handled.

**4.** Add a regular expression data type (Aho, Hopcroft, and Ullman 1974, pp. 318-361; Bourne 1983, pp. 185-186). Modify the functions find and match to operate appropriately when their first argument is a regular expression. Consider other situations in which regular expressions might be useful, as well as the possibility of operations on regular expressions.

5. Unify string and file data types so that string operations can be performed on files.

6. Implement lazy concatenation, in which strings are not actually concatenated until necessary but instead are kept in a list or similar structure. Decide whether this should be a special feature apparent to the source-language programmer or whether it should be a general property of the implementation of strings. Be careful to consider where type conversion and error checking should be performed. Include instrumentation to measure the performance of this feature.

7. Introduce the concept of strings of infinite length and provide ways of producing such strings. Integrate these strings into the regular string-processing repertoire of Icon.

8. Many of the useful properties of strings derive from the fact that they are linear sequences of characters. However, two-dimensional "strings" provide a useful representation of objects such as a printed page or a terminal display. A third dimension can be used for overlays. An extension of SNOBOL4 (Gimpel 1972b) provided a "block" data type for manipulating three-dimensional strings. Design and implement such a facility for Icon.

## E.3 STRUCTURES

The various structures in Icon are provided for handling aggregates of values that are organized in different ways. There are many possible extensions to operations on these structures, as well as other kinds of structures for handling different organizations of data. Here are a few possibilities:

1. Extend copy(x) for structures so that it makes a complete, physically distinct copy of x, rather than just a "top-level" copy. Consider, also, whether it would be better to provide a different function for this purpose and leave the present copy as it is.

2. Provide a function equiv(x, y) that succeeds if and only if x and y are structurally the same, not just identical objects as in x === y.

3. Add facilities for insertion and deletion of elements in the middle of lists, not just at the ends. Select a new data structure for representing lists internally so that these operations, as well as existing ones, can be performed efficiently.

4. Add a multi-dimensional array data type. Provide for lower bounds that may be different from 1.

5. Add a sparse array data type that efficiently implements arrays in which only a few elements have significant values.

6. Divide lists into two types: those that are fixed in size and those whose sizes may change by stack and queue access. Aim for an implementation that handles the two types most efficiently. Design the facility in a way that has the least impact on existing programs. Provide for conversion between the two types of lists.

7. Extend ''scalar'' operations, such as i + j, to operate element-wise on structures. For example, if a1 and a2 are lists, a1 + a2 should produce a new list consisting of the sums of the corresponding elements of a1 and a2.

8. Since both strings and lists are linear sequences, it is natural to have corresponding operations on the two types. A better solution might be to unify the two types, so that, for example, if a is a list, reverse(a) produces a list containing the elements of a in reverse order. Consider the repertoire of string-processing operations and how they could logically be extended to list-processing operations. Give careful consideration to conversion between strings and lists.

9. Add a function that deletes an element from a table.

10. Add a function that generates the entry values of a table.

11. Implement multisets, which are collections of values in which a value can occur more than once (Knuth 1973, pp. 22-34). Adapt set operations to apply to multisets.

12. Unify sets and csets by providing the operations of membership testing, insertion, and deletion for csets.

## 5.4 SCANNING

String scanning is one of the most powerful and useful features of Icon. Here are some suggestions for extending it:

1. Many problems that should be easy to deal with using string scanning are complicated by the fact that the data to be processed is contained in a file that consists of a sequence of lines that logically constitute a single string. To overcome this problem, implement a mechanism for automatically extending the value of &subject in the case that a matching function otherwise would fail because &subject is not long enough. Do this by providing a function to produce the extension. This function may, of course, fail.

2. Extend string scanning to apply to files.

3. Provide a facility for scanning lists and other structures in a fashion similar to string scanning. Consider new scanning state variables that may be needed because the elements of lists may be lists.

4. In the spirit of unifying string and list processing as suggested in a previous project, unify string and list scanning, using the result of the previous project as a starting point.

5. SNOBOL4 has patterns that contain matching procedures. Since patterns are data objects, they can be assigned to variables, passed to procedures, and so on. One of the advantages of patterns over matching expressions is that they provide a concise, declarative, high-level characterization of the properties of strings. Add patterns to Icon. Use SNOBOL4 patterns as a guide (Griswold, Poage, and Polonsky, 1971), but be careful to note where differences between the two languages should be reflected in how patterns are handled.

6. Add scanning (or pattern matching) for the two- and three-dimensional strings suggested in a previous project.

## E.5 FUNCTIONS, PROCEDURES, AND OPERATIONS

There are a number of "missing features" in the way Icon handles functions, procedures, and operators. The following projects provide extensions to fill the gaps:

1. Modify the function invocation mechanism so that is is possible to pass a variable to a function without its being dereferenced.

2. Add call-by-reference for procedures.

3. Provide a mechanism whereby a procedure can be called with an arbitrary number of arguments. Include a way for a procedure to determine the number of arguments with which it is called.
4. Generalize the handling of operators so that their meaning can be changed at run time.

5. Provide a way to declare new operators in a fashion similar to the way that procedures are declared.

6. At the present time, functions and procedures are called only by the interpreter. Provide a way whereby a function or procedure can be called from any C routine.

## E.6 TYPES

Icon has an unusually large number of data types, all of which are handled at run time. Several of the previous projects suggest additions of new types. Here are a few projects specifically related to types:

1. Add a new keyword &top, whose value is a new type, top, that is convertible to any type. For example, string(&top) might produce the empty string, while integer(&top) might produce zero. This new type provides the basis for a type hierarchy that has the form

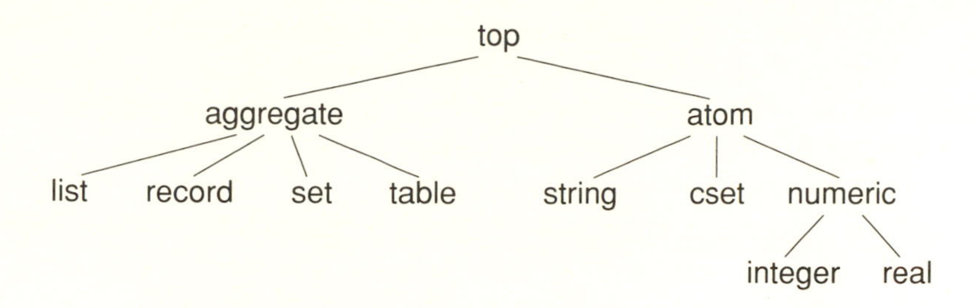


Fill out this hierarchy for the rest of the data types in Icon, being careful to place null appropriately. Add type-conversion functions for all types, including the classes aggregate and atom. Consider the implications of making top instead of null the initial value of all variables.

2. Design and implement facilities so that strong type checking is done when an Icon program is translated, rather than at run time. Give careful consideration to the problem of heterogeneity of aggregates. Design the facility in such a way that it has the minimum impact on existing programs.

## E.7 TRAPPED VARIABLES

Sometimes an implementation technique suggests a language feature. The concept of trapped variables was introduced in the SITBOL implementation of SNOBOL4 as a mechanism for handling implicit actions that may occur as side effects of dereferencing or assigning to a variable (Gimpel 1972a). For example, in SNOBOL4

```
OUTPUT = S
```

not only assigns the value of S to OUTPUT, but it also writes that value to standard output. Similarly,

```
S = INPUT
```

reads a line from standard input and assigns it to S. If an end of file is encountered, the value of S is not changed and the assignment statement fails in a fashion similar to the failure of read() in Icon.

The merit of the trapped-variable mechanism used in SITBOL lies partly in its generality. The ''normal'' value of a variable is replaced by a pointer to a trapped-variable block that hides this ''normal'' value. The trapped-variable block also contains the information necessary to carry out appropriate actions if the variable is assigned to or dereferenced (Hanson 1976).

1. Design and implement a uniform approach to trapped variables in Icon so that keyword trapped variables, substring trapped variables, and table-element trapped variables all have the same structure. Consider the effect of such a change on the amount of storage needed for trapped variables and the time required to process keyword, substring, and table references.

2. In SNOBOL4, operations associated with dereferencing a trapped variable may fail. In Icon, however, failure cannot occur when dereferencing a trapped variable. Investigate the implications of possible failure during the dereferencing of trapped variables in Icon and modify the implementation as necessary so that this possibility is handled properly.

3. Implement input and output associations for Icon in the style of SNOBOL4. Add trace associations as a generalization of input and output associations.

4. Add programmer-defined associations to Icon so that a procedure can be invoked when dereferencing or assignment occurs.

5. Design and implement other types of associations, such as read-only protection of variables and the monitoring of events associated with global and static identifiers.

## E.8 STREAMS

Co-expressions provide a general mechanism for producing results at the time and place that they are needed, but a much simpler mechanism would suffice for producing the elements of data objects in sequence. All that is needed is a reference to the object and an index that keeps track of a position in the object.

1. Implement a new data type stream and a function stream(x) that produces a stream from which the elements of x can be produced. Provide a function Next(s) that produces the next element from the stream s.

2. Add a function Current(s) that produces the last element produced by Next(s) and a function Value(s) that produces the value from which s produces elements. Modify the size operation so that *s produces the number of elements that a stream s has produced.

3. Describe what would be involved in implementing a function Previous(s) that produces the previous element in the stream and decrements the index.

## E.9 PARALLEL EVALUATION

The order in which results are produced by generators in Icon is motivated by searching for solutions among all possibilities and amounts to the depth-first traversal of "solution trees." On the other hand, there are situations in which the results of generators are needed in parallel. For example, in numerical computation, it might be useful to have !a1 + !a2 generate the sums of corresponding values in the arrays a1 and a2.

Such evaluation mechanisms can be constructed using co-expressions (Novak and Griswold 1982; Griswold and Novak 1983), but that approach is both inefficient and cumbersome.

1. Design a mechanism for producing the results of generators in parallel or according to similar "evaluation regimes." Consider whether such a mechanism should be a property of the operation on generators or the context in which the operation appears.

2. Provide programmer-defined evaluation regimes.

*Comment:* These projects are, of course, *very* difficult if carried out in the general context of the current implementation of Icon, in which the last-in, first-out order of generation is fundamental.

## E.10 DIAGNOSTIC FACILITIES

Icon's diagnostic facilities are comparatively weak. Here are some suggestions for improvement:

1. Provide a way to trace only a single procedure or a set of procedures, as opposed to every procedure.

2. Provide a facility for tracing functions.

3. Provide a "trace-back" facility, so that in the case of a run-time error, the expression in which the error occurred is printed, followed by the series of procedure calls that led to it.

4. Provide a facility for tracing the usage of a variable so that trace output occurs when the variable is dereferenced or when a value is assigned to it. Take into account keywords such as &random.

5. Provide a tracing facility for string scanning that produces trace output when &pos or &subject is changed. Provide enough context to make the output understandable in the context of string scanning.

6. Provide a facility for tracing co-expression activity.

## E.11 A RECURSIVE INTERPRETER

The expression, generator, and procedure frames that the interpreter pushes on the interpreter stack can be viewed as storing information that otherwise would be placed on the C stack as the normal result of a call if the interpreter were called in such situations. That is, the i-state variables that are saved on the interpreter stack might, instead, be arguments or local variables of the interpreter procedure.

Design and implement a ''recursive'' interpreter in which the interpreter stack contains only descriptors and in which frame information is handled by recursive calls of the interpreter.

*Comments:* This project is more difficult than it may seem at first sight. It requires a deep understanding of expression evaluation in Icon and what goes on in the interpretive process. In designing a recursive interpreter, it is important to determine what constitutes the state of the interpreter and what arguments and local variables the interpreter should have. The number of arguments and local variables should be small to avoid excessive stack depth. Some control structures present special problems, yet the interpreter should not be burdened with arguments or local variables that are needed only for specific control structures.

One way to approach a problem like this is to construct a prototype implementation in a high-level language that allows more flexibility than the eventual implementation language. For this project, Icon provides a natural prototyping language for the eventual C implementation.

## E.12 INTERACTIVE FACILITIES

Icon is not, itself, an inherently interactive language. Although a program written in Icon can provide an interactive user interface, it cannot modify or extend itself. The ultimate project in this area is the design and implementation of a programming environment for Icon. Here are a few suggestions for starting points:

1. Provide a way to load functions (written in C) at run time.

2. Provide a way to load procedures (via icode files) at run time.

3. Provide a way to create procedures at run time.

4. Provide a way for the core image of a running program to be dumped to a file so that it can be loaded to restart the program at a later time. Note that such a feature is inherently specific to the operating system under which Icon runs.

5. Design and implement a terminal-independent windowing package for Icon.

6. Design and implement a symbolic debugger for Icon. Give special consideration to generators.

## E.13 INSTRUMENTATION

Icon is so different from most other programming languages that it is difficult to measure the performance of its implementation or even to determine the most efficient way of performing a computation.

There is some instrumentation in Icon to help gather data that might be useful for making such measurements, but it is relatively rudimentary. Here are some suggestions for more sophisticated facilities:

1. Design and implement a benchmarking facility that can be used to determine how much time it takes to evaluate a specific Icon expression. Be careful to consider the possible effects of garbage collection. Use this benchmarking facility to develop guidelines for Icon programmers and to search for possible inefficiencies in the implementation.

2. Instrument the interpreter to count the number of times each different virtual machine instruction is executed. Collect the results in an array and print them out at the end of program execution, provided an appropriate environment variable is set.

3. Make a catalog of the most significant operations, both at the source-language level and in the implementation itself, that would contribute to an understanding of the behavior of Icon programs and the performance of its implementation. Examples are type conversions, including those that prove unnecessary, storage allocation, garbage collection, and so on. Devise a framework for collecting information on these operations and reporting it in a useful way.

**4.** Instrument the interpreter to report, at the end of program execution, the high-water marks for the following:

- interpreter stack (sp)
- procedure call level (&level)
- nesting depth of expression frames
- nesting depth of generator frames
- interpreter level (ilevel)

**5.** Instrument the interpreter to dump the values of i-state variables and the contents of the interpreter stack. Provide a way whereby this information can be obtained for different degrees of execution "granularity," such as for the execution of every virtual machine instruction or only when a specific virtual machine instruction is executed. Write a program (in Icon) to convert the data obtained in this way into pictures of the interpreter stack, such as those contained in Chapters 8, 9, and 10 of this book.

**6.** Instrument the hashing routine and devise a method for measuring the performance of hashing for values of various types.

**7.** Instrument set and table lookup to measure its performance for various types of data.

**8.** Instrument Icon's storage-management system to report the details of storage allocation and garbage collection. Include such information as the size of the list of pointers to qualifiers (quallist), the depth of recursion in marking blocks, and the effectiveness of the breathing-room heuristics.

**9.** An unconventional but powerful tool for measuring some aspects of program behavior consists of source-language functions that allow access to internal information at run time (Griswold 1975; Griswold 1980a). With such a tool, an "introspective" program can be written to monitor its own activity. Implement the following functions:

| | |
|---|---|
| Dword(x) | produce the d-word of x as an integer |
| Vword(x) | produce the v-word of x as an integer |
| Descr(x, y) | construct a descriptor using x as the d-word and y as the v-word |
| Indir(x) | produce the descriptor at the location x |
| Ivar(s) | produce the value of the i-state variable whose name is s (for example, Ivar("ilevel") should produce the value of ilevel) |

Use these functions to write Icon programs to determine such things as the lengths of the chains in sets and tables.

**10.** Provide source-level profiling for Icon programs.

APPENDIX F

# Solutions to Selected Exercises

**2.2** The expression

```
write("hello" | "howdy")
```

just writes hello; there is no context to cause the second argument of the alternation to be produced.

**2.3** The expression

```
|(1 to 3) > 10
```

is an evaluation "black hole." The left argument produces 1, 2, 3, 1, 2, 3, 1, 2, 3, ... indefinitely, since the comparison always fails.

**3.1** The difference in execution time between interpretation and the execution of compiled code is comparatively small, since most of the time in execution is spent in subroutines that are called in either case.

**3.3** The lexical analyzer inserts a semicolon between the lines

```
s1 := s2
|| s3
```

because an identifier is legal at the end of an expression, and the unary operator for repeated alternation is legal at the beginning of an expression. Thus, the second line consists of two applications of repeated alternation to s3.

**4.6** Type checking is a byproduct of conversion. If conversion were not automatic, type checking would be simple, but the conversion routines would still be needed for explicit type conversion. Without implicit type conversion, the programmer would have to perform explicit conversions, as in

```
while sum +:= numeric(read())
```

However, since both read and numeric can fail, a more complicated construction would be advisable:

```
while x := read() do
   sum +:= numeric(x) | stop("nonnumeric data")
```

This illustrates why the failure of an implicit conversion is an error.

**4.7** A test is necessary to distinguish between qualifiers and other descriptors, regardless of which has the flags. However, if qualifiers had flags, it would be necessary to mask them out in order to extract their lengths. Since masking is needed anyway to extract the type codes of other descriptors, it is more efficient to relegate the flags to them.

**4.9** The basic approach to designing one-word descriptors is to have them point to intermediate data that otherwise would be contained in two-word descriptors. In the simplest form, one-word descriptors would point to two-word blocks corresponding to two-word descriptors (Hanson 1980).

**5.1** If Icon is implemented on a computer whose collating sequence is different from ASCII, such as EBCDIC, it still uses the ASCII collating sequence for comparison and sorting. Consequently, these operations may produce results that are different from ones that are expected in an EBCDIC environment. If the native character set of a computer is smaller than 256, there may be problems implementing Icon, since the size of a C char may be less than expected in some routines.

**5.4** Even if a newly created string exists in string space, the code to find it would be complex and the time required probably would outweigh any advantages of avoiding duplicate allocation. Furthermore, the new string must be constructed somewhere first in order to know what to look for. This construction often is most conveniently performed in allocated string space, so most of the work is done already.

**5.7** If the context in which a subscripting expression is used can be determined by the translator, different virtual machine instructions could be generated for the different contexts. If the context is dereferencing, a virtual machine instruction for producing a substring could be used to avoid the construction of a substring trapped variable and the allocation of space that might have to be reclaimed during garbage collection. Similarly, if the context is assignment, virtual machine instructions for the appropriate concatenation could be produced.

**5.10** The following procedure illustrates the usefulness of polymorphic subscripting:

```
procedure shuffle(x)
   every !x :=: ?x
   return x
end
```

This procedure can shuffle strings, lists, and records. Note that subscripting is implicit in the element-generation and random-selection operations. The same principle applies to explicit subscripting, however.

**6.2** An empty list consists of a list-header block and one list-element block. For convenience in diagramming, the list-element blocks in Chapter 6 have space for only four elements, but the normal number is eight. Therefore, on a computer with 32-bit words, a list-element block occupies 100 bytes, while a list-header block occupies 24 bytes. Thus, on a computer with 32-bit words, an empty list occupies 124 bytes in all. On a computer with 16-bit words, the empty list is half this big. This may seem like a lot of space for ''nothing,'' but an empty list usually is created so that elements can be added to it, in which case the overhead becomes progressively less significant.

**6.4** Just because a list element is removed from a list by a stack or queue operation does not mean that the element is inaccessible. Consider the expression

```
process(a[1], pop(a))
```

Since the arguments are not dereferenced until both of them are evaluated, it might prove surprising if the value of a[1] were changed by the pop. Although such situations are unlikely in practice, it is important that list accesses behave coherently and consistently. One way to view the handling of this situation is that the pop does not remove an element but only makes it unavailable for *subsequent* access.

**6.5** As indicated by the preceding solution, references to list elements that are removed by stack or queue operations may remain. It is easy to construct examples in which there are such references to unlinked list-element blocks.

**7.3** An empty set consists of just a set-header block. There are two words for the title and size. A computer with 16-bit words has thirteen 4-byte slots, while a computer with 32-bit words has thirty-seven 8-byte slots. Thus, on a computer with 16-bit words, an empty set occupies 56 bytes, while on a computer with 32-bit words, an empty set occupies 304 bytes. Each member added to a set occupies 12 bytes on a computer with 16-bit words and 24 bytes on a computer with 32-bit words.

**7.8** It is necessary to access a table to determine whether an element is already in it. Since the table contains the default value, it is readily available when it is needed. Because the default value may not be needed, it is not worth the trouble always to put it in table-element trapped-variable blocks.

**7.12** The default value of a table is constant and could be used for computing its hash number.

**7.13** If hashing is based on an attribute of a value that can change, two hash computations on that value may produce different results. If this happens, the value appears to be different in the two cases. Thus, the same value might be inserted in a set twice, an element that is in a table might not be found, and so on.

**8.1** The str instruction pushes a descriptor on the interpreter stack. The d-word of this descriptor is the string length, and the v-word is a pointer to its location. Since the interpreter stack grows toward increasing memory addresses, it is natural and convenient to push the v-word first and to have its value as the first operand of the str instruction.

**8.3** The two expressions

```
a[?i] := a[?i] + 1
```

and

```
a[?i] +:= 1
```

generally produce different results.

**8.4** A new operator requires new syntax, which must be handled in the translator. Since each operator has its own virtual machine instruction, a new instruction must be added for the new operator. The translator, linker, and run-time system all must handle this new virtual machine instruction. In the run-time system, the code for this instruction must be added to the main switch statement in the interpreter, with an appropriate call to a C routine to implement the operation.

**9.2** Suspension of functions and operations (but not procedures) increases the interpreter level. This occurs both in nested generators and generators in mutual evaluation. Thus, if $expr_1$ is a function or operator that suspends, the interpreter level is increased by its evaluation in both

*$expr_2$*(*$expr_1$*)

and

*$expr_1$* & *$expr_2$*

**9.5** Exercise 2.3 contains an example of a "black hole" as a result of repeated alternation.

**9.7** If the result of *$expr_1$* in

every *$expr_1$* do *$expr_2$*

is not popped, the interpreter stack builds up for every value produced by *$expr_1$*. While these descriptors are removed when the loop terminates, interpreter stack overflow would result in expressions like

```
every 1 to 100000 do
   expr2
```

**9.8** The expressions

every *$expr_1$*

and

*$expr_1$* & &fail

are equivalent unless *$expr_1$* contains a break or next expression.

**10.1** If a procedure or function call contains an extra argument that fails, the call is not performed (more accurately, the previous argument is resumed).

**10.3** If a variable whose value is on the interpreter stack is returned from a procedure, the variable must be dereferenced, since the value on the stack may be overwritten. Thus, arguments and dynamic local identifiers are dereferenced on procedure return. Static local identifiers are not dereferenced, since they are not on the stack. There is a good argument why static local identifiers should be dereferenced, since local identifiers are expected to be accessible only within the procedures in which they are declared. Suspension also dereferences variables whose values are on the stack. These values will not be overwritten, however, since the portion of the stack containing these values is not removed until the surrounding expression frame is removed, at which time the variables in it are no longer accessible. The rationale for dereferencing on suspension is that there should be no difference, at the source-language call, between a procedure that returns and a procedure that suspends. Furthermore, if dereferencing on suspension were not done, the arguments and local identifiers of the suspending procedure could be modified by the procedure that calls it. This would violate the semantic concept of "local," although, as mentioned previously, static local identifiers are not dereferenced. This may be viewed as a bug.

**10.4** The interpreter stack grows upward, regardless of the architecture of the computer on which Icon is implemented. On a computer with an upward-growing C stack, the C stack base is in the middle of the co-expression block, instead of at its end. There is an advantage, in co-expressions, to having a downward-growing C stack: the interpreter and C stacks grow toward each other, sharing space, and run out of space only when they collide. For an upward-growing C stack, either the interpreter stack or the C stack may run out of space, even if there is unused space in the other.

**11.1** Having the type code in the d-word of a descriptor allows its type to be determined without a level of indirection to the block title through its v-word.

**11.2** The keyword trapped-variable block for &subject is statically allocated in the run-time system, so any scanning operation or assignment to &subject changes the contents of a block in the statically allocated portion of the run-time system.

**11.3** The values of global and static identifiers are in the icode region, so any assignment to one of them changes data in the icode region.

**11.4** Global and static identifiers could be in a single array in the icode region. It is simply more convenient to handle them separately in the translator and linker.

**11.5** The names of global identifiers are needed for the display function and for the string invocation of functions as described in Exercise 10.2.

**11.6** The names of static identifiers are in their procedure blocks.

**11.17** The maximum type code is a small integer. Therefore, it is only necessary that the allocated block region be at an address larger than this value. This happens automatically because of the way the code for the run-time system is arranged, although it would be trivially easy to force.

**11.19** If there were more than one pointer on quallist to the same qualifier, the v-word of that qualifier would be relocated more than once, with an erroneous result. Such an error would show up when the qualifier was used in subsequent program execution after garbage collection.

**11.21** The keyword trapped-variable block for &subject may contain a pointer to the allocated string region. This pointer is processed because the keyword trapped-variable block for &subject is in the basis. Other such cases could be handled the same way.

**11.26** Recursion in markblock occurs when a block contains a pointer to another block. Thus, pointer chains cause recursion of corresponding depth. The following program builds a linked list whose length is the number of lines in the input file:

```
record lines(value, nextl)

procedure main()
   head := lines()
   current := head
   while current.value := read() do {
      current.nextl := lines()
      current := current.nextl
      }
        .
        .
        .
end
```

In the case of a garbage collection, the depth of recursion in markblock is the length of the linked list.

**11.31** An excessively large predictive need request may trigger a garbage collection unnecessarily or possibly cause error termination if there is not enough memory, although this is unlikely unless the request is enormous. If the request is satisfied, execution continues normally. If a predictive need request is less than the amount subsequently allocated *and* there is not enough space in the region to satisfy the allocation request, the allocation routine terminates program execution with a message indicating that there is an error in the implementation of Icon.

**11.32** As indicated in the solution to Exercise 6.5, an unlinked list-element block may still be referenced by a variable. As long as this is the case, the list-element block cannot be reclaimed by garbage collection. In general, as long as there is a pointer to a block that is reachable from the basis, the block cannot be reclaimed.

**11.33** A variable could point to the head of the block, with an offset referred to the location of the corresponding value. The advantage of having a variable point directly to its value is efficiency during expression evaluation. Access to variables in blocks is presumably more frequent during expression evaluation than during garbage collection. There can be at most one access to such a variable during a garbage collection, while the variable may be accessed repeatedly during expression evaluation. Furthermore, many programs use such variables frequently but never require garbage collection.

**11.38** In the case of a type, such as real, for which all blocks are the same size, a separate allocation region can be efficiently managed by a free-list mechanism. No relocation is needed in such a region, and since all blocks are of the same size, there is no problem with fragmentation. Having many allocation regions complicates the storage-management system, of course. The implementation of an early version of Icon employed several regions in which the space for freed blocks was reused (Hanson 1980).

**11.39** It is not possible to determine, in general, if storage is needed only temporarily or if it must be retained for an arbitrarily long time. For example, in

```
while process(read())
```

the retention of storage for input strings depends entirely on what process does. If it just writes out its argument, as shown in the exercise, storage for the strings produced by read would not have to be allocated. If, on the other hand, it pushes its argument on a list, the strings must be allocated. Since allocation is fast and garbage collection has little work to do for space that is not accessible, storage throughput has a comparatively small impact on performance.

**12.2** On computers with 16-bit words, tests for the two types of integers must be made in all situations where type checking and conversion are performed. Furthermore, all computations that may produce a source-language integer that requires more than 16 bits must provide for the creation of long-integer blocks. Since integers occur in many contexts, a substantial amount of code is required to handle the two types of integers.

**12.4** The value of MaxCvtLen must be sufficient to handle a string whose length is the size of the character set. Thus, if Icon had 512 different characters, MaxCvtLen would have to be 512. Even if the number of different characters were small, it is unlikely that some other type of conversion, such as real-to-string, would dominate.

**12.14** In write(x1, x2, ..., xn), the arguments are converted to strings, if necessary, for the purposes of output. With the exception of xn, such strings resulting from conversion do not have to be placed in allocated storage, since they are used only transiently. However, the value returned by write is the string value of xn, so if it is not a string, its converted value must be placed in allocated string storage.

# REFERENCES


Aho, Alfred V.; Hopcroft, John E.; and Ullman, Jeffrey D. 1974. *The design and analysis of computer algorithms*. Reading, Massachusetts: Addison-Wesley Publishing Company.

Aho, Alfred V.; Sethi, Ravi; and Ullman, Jeffrey D. 1985. *Compilers: principles, techniques, and tools*. Reading, Massachusetts: Addison-Wesley Publishing Company.

Bourne, S. R. 1983. *The UNIX system*. London: Addison-Wesley Publishing Company.

Clocksin, W. F., and Mellish, C. S. 1981. *Programming in Prolog*. New York: Springer-Verlag.

Cohen, Jacques. 1981. Garbage collection of linked data structures. *Computing Surveys*. 13:341-367.

Dewar, Robert B. K., and McCann, Anthony P. 1977. Macro SPITBOL—a SNOBOL4 compiler. *Software—Practice and Experience*. 7:95-113.

Dewar, Robert B. K.; Schonberg, E.; and Schwartz, J. T. 1981. *Higher-level programming; introduction to the use of the set-theoretic programming language SETL*. New York: Computer Science Department, Courant Institute of Mathematics, New York University.

Farber, David J.; Griswold, Ralph E.; and Polonsky, Ivan P. 1964. SNOBOL, a string manipulation language. *Journal of the ACM*. 11,1:21-30.

Farber, David J.; Griswold, Ralph E.; and Polonsky, Ivan P. 1966. The SNOBOL3 programming language. *The Bell System Technical Journal*. XLV:895-944.

Gimpel, James F. 1972a. SITBOL; Version 1.0. Murray Hill, New Jersey: Bell Laboratories Technical Report S4D30.

Gimpel, James F. 1972b. Blocks—a new datatype for SNOBOL4. *Communications of the ACM*. 15:438-477.

Gonnet, G. H. 1984. *Handbook of algorithms and data structures*. London: Addison-Wesley Publishing Company.

Griswold, Ralph E. 1972. *The macro implementation of SNOBOL4; a case study of machine independent software development*. San Francisco: W. H. Freeman and Company.

Griswold, Ralph E. 1975. A portable diagnostic facility for SNOBOL4. *Software—Practice and Experience*. 5:93-104.

Griswold, Ralph E. 1977. The macro implementation of SNOBOL4. In *Software portability*, ed. P. J. Brown, pp. 180-191. Cambridge: Cambridge University Press.

Griswold, Ralph E. 1980a. Linguistic extension of abstract machine modelling to aid software development. *Software—Practice and Experience*. 10:1-9.

Griswold, Ralph E. 1980b. The use of character sets and character mappings in Icon. *The Computer Journal*. 23, 2:107-114.

Griswold, Ralph E. 1981. History of the SNOBOL languages. In *History of programming languages*, ed. R. L. Wexelblat, pp. 601-660. New York: Academic Press.

Griswold, Ralph E., and Griswold, Madge T. 1983. *The Icon programming language*. Englewood Cliffs, New Jersey: Prentice-Hall, Inc.

Griswold, Ralph E., and Hanson, David R. 1977. An overview of SL5. *SIGPLAN Notices*. 12,4:40-50.

Griswold, Ralph E., and Hanson, David R. 1979. *Reference manual for the Icon programming language*. Tucson, Arizona: Department of Computer Science Technical Report TR 79-1, The University of Arizona.

Griswold, Ralph E.; Mitchell, William H.; and O'Bagy, Janalee. 1986. *Version 6.0 of Icon*. Tucson, Arizona: Department of Computer Science Technical Report TR 86-10, The University of Arizona.

Griswold, Ralph E., and Novak, Michael. 1983. Programmer-defined control operations in Icon, *The Computer Journal*. 26, 2:175-183.

Griswold, R. E.; Poage, J. F.; and Polonsky, I. P. 1971. *The SNOBOL4 programming language*. Englewood Cliffs, New Jersey: Prentice-Hall, Inc. 2nd edition.

Hanson, David R. 1976. Variable associations in SNOBOL4. *Software—Practice and Experience*. 6:245-254.

Hanson, David R. 1977. Storage management for an implementation of SNOBOL4. *Software — Practice and Experience*. 7:179-192.

Hanson, David R. 1980. A portable storage management system for the Icon Programming Language. *Software—Practice and Experience*. 10:489-500.

Hanson, David R., and Griswold, Ralph E. 1978. The SL5 procedure mechanism. *Communications of the ACM*. 21:392-400.

Johnson, Stephen C. 1975. *Yacc— yet another compiler-compiler*. Murray Hill, New Jersey: Bell Laboratories Computing Science Technical Report 32.

Kernighan, Brian W. 1975. RATFOR—a preprocessor for a rational Fortran. *Software—Practice and Experience*. 5:395-406.

Kernighan, Brian W., and Ritchie, Dennis M. 1978. *The C Programming Language*. Englewood Cliffs, New Jersey: Prentice-Hall, Inc.

Knuth, Donald E. 1969. *The art of computer programming*. Seminumerical algorithms, vol. 2. Reading, Massachusetts: Addison-Wesley Publishing Company.

Knuth, Donald E. 1973. *The art of computer programming*. Sorting and searching, vol. 3. Reading, Massachusetts: Addison-Wesley Publishing Company.

Lesk, M. E. 1975. *Lex—a lexical analyzer generator*. Murray Hill, New Jersey: Bell Laboratories Computing Science Technical Report 39.

Liskov, Barbara. 1981. *CLU reference manual*. New York: Springer-Verlag.

McKeeman, W. M.; Horning, J. J.; and Wortman, D. B. 1970. *A compiler generator*. Englewood Cliffs, New Jersey: Prentice-Hall, Inc.

Newell, Allen. 1961. *Information Processing Language-V manual*. Englewood Cliffs, New Jersey: Prentice-Hall, Inc.

Newey, M. C.; Poole, P. C.; and Waite, W. M. 1972. Abstract machine modelling to produce portable software. *Software—Practice and Experience*. 2:107-136.

Novak, Michael, and Griswold, Ralph E. 1982. *Programmer-defined evaluation regimes*. Tucson, Arizona: Department of Computer Science Technical Report TR 82-16, The University of Arizona.

Reiser, John F., ed. 1976. *SAIL*. Stanford, California: Stanford Artificial Intelligence Memo AIM-289, Computer Science Department Report STAN-CS-76-574.

Ritchie, D. M., and Thompson, K. 1978. The UNIX time-sharing system. *The Bell System Technical Journal*. 57:1905-1930.

Santos, Paul Joseph, Jr. 1971. *FASBOL, a SNOBOL4 compiler*. Berkeley: Electronics Research Laboratory, University of California Memorandum ERL-M134.

Shaw, Mary. 1981. *Alphard: form and content*. New York: Springer-Verlag.

Teitelman, Warren. 1974. *InterLisp reference manual*. Palo Alto, California: XEROX Palo Alto Research Center.

Waite, William M., and Goos, Gerhard. 1984. *Compiler construction*. New York: Springer-Verlag.

Yngve, Victor H. 1960. The COMIT system for mechanical translation. In *Information Processing: Proceedings of the International Conference on Information Processing*. Munich: R. Oldenbourg.

# INDEX

Library of Congress Cataloging-in-Publication Data

Griswold, Ralph E., 1934-
The implementation of the Icon programming language.

(Princeton series in computer science)
Bibliography: p.
Includes index.
1. Icon (Computer program language) I. Griswold,
Madge T., 1941- II. Title III. Series.
QA76.73.I19G76 1986 005.12'3 86-42844
ISBN 0-691-08431-9 (alk. paper)